BATTLE READY

BATTLE READY

The National Coast Defense System and the Fortification of Puget Sound, 1894–1925

DAVID M. HANSEN

Washington State University Press
Pullman, Washington

Washington State University Press
PO Box 645910
Pullman, Washington 99164-5910
Phone: 800-354-7360
Fax: 509-335-8568
Email: wsupress@wsu.edu
Website: wsupress.wsu.edu

Library of Congress Cataloging-in-Publication Data

Hansen, David M.
  Battle Ready : the National Coast Defense System and the fortification of Puget Sound, 1894-1925 / David M. Hansen.
    pages cm
  Includes bibliographical references and index.
  ISBN 978-0-87422-320-0 (alk. paper)
1. Coast defenses—Washington (State)—Puget Sound—History. 2. Fortification—Washington (State)—Puget Sound—History. 3. Military architecture—Washington (State)—Puget Sound 4. Puget Sound (Wash.)—History, Naval. 5. Puget Sound (Wash.)—History, Military.
I. Title. II. Title: National Coast Defense System and the fortification of Puget Sound, 1894-1925.
  UG412.P85H35 2014
  355.4'509797709041—dc23
                                                                                           2014004600

Fine Quality Books from the Pacific Northwest

For my dad, Alven Christian Hansen

Table of Contents

List of Plates

Preface

I came across a coastal fortification for the first time when I was not even old enough to drive. I wasn't sure what it was, but I knew that it was unlike any other kind of building that I had ever seen, and I knew that I wanted to know more about it. As it sometimes happens (or at least as we sometimes like to remember), an early experience can influence the rest of our lives, and for me that single encounter led to a career in architectural history and a specific interest in military architecture of more than fifty years running.

The Puget Sound location was a natural choice—it was where I lived—and also a happy one since much of the research on its history and construction could be carried out locally, and as it turned out, the works in Puget Sound were among the most challenging in the national system. It was an ideal instance in a larger context.

The present book is an effort to help others who might be mystified and intrigued by the scattered collection of concrete monoliths that form the most memorable elements of the defenses.

There are two stories here. One is obvious: the design, construction, and use of the fortifications and their related parts. Put another way, it is the history of what we see when we visit these places today. The other is discreet: the experiences of those directly involved with building the defenses and then figuring out how to employ them.

The first story is the easiest to tell. Thanks to the diligent record-keeping of the military and the technical literature of the time, we know just about all we could wish to know about the chronicle of events and dates concerning the transfiguration of peaceful shorelines into expressions of national defense. It is a subject that is awash in minutiae. It has been frustrating to face the skimpy documentation for the second tale. Historians rely on letters, diaries, personnel papers, and other sources of recollection to add dimension to the otherwise invisible human narrative of motivation and perspective that every good story contains, yet those materials have been especially hard to find. We know the names of the principal characters and many of the supporting players, and after looking long and hard in the places where their written legacies may have come to rest, we have to conclude that almost to a man they have left us with only scattered hints to suggest what they might have thought about what was happening around them. It doesn't mean that those valuable papers do not exist; we can hope that they survive and have yet to surface. When they do materialize, there will be a welcome opportunity to add to the present outline.

I have had much help over the years and specifically with the manuscript of this book. Thanks are due to Greg Hagge, at the time of this writing the curator of arms and armor at the U.S. Army Ordnance Training and Heritage Center, who gave a detailed examination of an early draft and made many helpful suggestions; and to Bolling Smith, editor of the *Coast Defense Study Group Journal*, for his comprehensive comments on the text. William Woodward, PhD, professor of history at Seattle Pacific University, shared his keen insights and his observations contributed substantially to the finished work. I am indebted as well to Steven Kobylk and Dan Kerlee for access to their photo collections, as well as to Ryan Karlson and Alicia Woods of the Washington State Parks and Recreation Commission for their help in using the historical materials in that agency's possession.

And finally, there are two individuals who deserve special mention. Without their help and friendship early on it is unlikely that I would have continued my interest. CDR David P. Kirchner, USN (Ret), was a companion and guide on many occasions and his generosity was the foundation of my own research. E. R. Lewis, author of the most useful work to date on the coastal fortifications of the United States, readily straightened out questions I thought hopelessly knotted and gave freely of his time and knowledge.

There are others unnamed who have helped write the book that I wished to have had with me on that first visit. I hope I have done them justice. It is a complex story with many threads and many opportunities for missteps; I believe that I have avoided most of them but no doubt there are errors to be discovered and for that I alone am responsible.

Introduction

These pages seek to illuminate the creation of the Coast Defenses of Puget Sound, a collection of five fortifications that formed one segment of a national system of protection against naval attack on important seacoast cities and harbors. It looks closely at the design and construction of the fortifications, and also the technology that helped bring them about and that was essential for their operation. It is a story that has a long beginning and a long conclusion, but in the main the emphasis is on the period between 1885 and 1924.

The selection of dates is in some ways artificial. The starting point of 1885 is reasonable since it was in that year that Congress authorized the creation of the national fortification plan that would ultimately (but not initially) include Puget Sound. The end date of 1924 has more to do with a change in perception of coast defense than it does with any single important event. However, bracketed by those years was a period of significant creativity and enthusiasm, expressed most visibly by the extensive construction effort that was followed by the energetic pursuit of methods to use and, later, improve what had been built. World War I had a deep impact on the coast artillery service as it had developed to that time, and those effects had become most apparent by the 1920s. Certainly the practice of coast defense continued on afterward, and saw additional construction during World War II, but it had none of the sense of newness that accompanied the growth years that began at the end of the nineteenth century.

Puget Sound is an important and instructive location for the study of the defenses writ large. Individuals who were stationed here were major contributors to the national system, either in the design of the fortifications or in the supervision of the posts once construction was complete. From the point of view of architectural history, the location presents an excellent opportunity to consider the evolution of fortification forms. It was also not an easy place to protect, and that difficulty gives us the opportunity to examine the ways in which the army's engineers sought to solve the problem. And finally, it is a system that is today exceptionally intact. Considering Fort Worden, Fort Casey, and Fort Flagler, all of the major components of the fortification scheme survive within a landscape that is almost unchanged, although the same cannot be said of the much smaller facilities at Fort Ward and Fort Whitman.

Several sources have been key, chief among them the records of the chief of engineers in the National Archives that detail the construction work in Puget Sound. Another important source has been the *Journal of the United States Artillery*, which published many articles by artillery officers that described the changes taking place in their service during the period covered. There seem to be no end of historic photographs, many taken by the soldiers themselves as hobbyists and later sold as souvenir postcards. The Corps of Engineers also photographed the progress, and an almost complete set of prints from the large-format glass negatives exists.

The material here is for a general audience, most likely those individuals who have had some contact with the former defenses in their current form as state parks. Keeping in mind that the technical detail that is fascinating for enthusiasts can also be a painful slog for most readers, many explanations have been streamlined for the sake of brevity and clarity. Fortunately for those who do have a deeper interest, there are organizations whose activities center on the historical technology of coast defense and whose members have marshaled an impressive understanding of the subject.

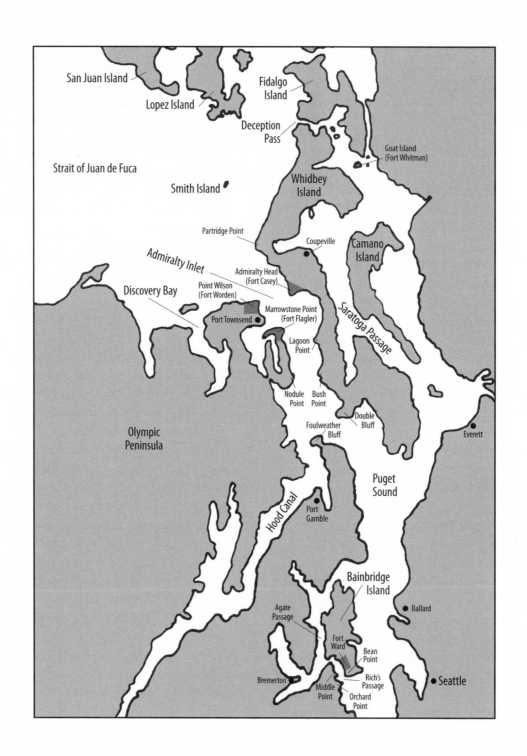

San Juan Island

Lopez Island

Fidalgo Island

Deception Pass

Strait of Juan de Fuca

Goat Island (Fort Whitman)

Smith Island

Whidbey Island

Partridge Point

Coupeville

Camano Island

Admiralty Inlet

Admiralty Head (Fort Casey)

Point Wilson (Fort Worden)

Discovery Bay

Marrowstone Point (Fort Flagler)

Port Townsend

Saratoga Passage

Lagoon Point

Nodule Point

Bush Point

Double Bluff

Foulweather Bluff

Olympic Peninsula

Everett

Puget Sound

Hood Canal

Port Gamble

Bainbridge Island

Agate Passage

Ballard

Fort Ward

Bean Point

Bremerton

Rich's Passage

Middle Point

Orchard Point

Seattle

The Wilderness of Waters

"Everything has yet to be done here. Years must be consumed
in the execution of any project or system, no matter what."
—*General J. G. Totten, 1860*

It had been a busy afternoon. The program, arranged by the military affairs committee of the Seattle Chamber of Commerce, had included a number of prominent military men. Also on the agenda was a representative of the chamber's own organization that was studying the maintenance of an army in the Seattle area should war come again. In the spring of 1926, as the audience listened to the messages from the platform, war seemed distant, a subject interesting primarily for its business opportunities. There was a luncheon, and as the guests and attendees were served, the Coast Artillery Band from Fort Worden furnished a concert.[1]

Fort Worden, near Port Townsend and overlooking Admiralty Inlet, was the headquarters of the system of permanent fortifications that protected Puget Sound. Traveling from the fort with the band was the current commander of the defenses, Colonel Percy M. Kessler. Kessler had been invited to speak to the chamber as part of its national defense program.

Kessler was a good choice. In so far as the country's commitment to coast defense went, Kessler had just about seen it all. He had graduated from the Military Academy at West Point in 1896 as a member of a military force in transformation. The Civil War was then a generation and more past and the army was sloughing off its post-war responsibilities of strikebreaking and acting as the constabulary in the former states of the Confederacy during Reconstruction. The long years of pursuit and skirmish with Indian tribes in the southwest, the Rocky Mountain west, and the Plains

was also at an end, concluding what had been the army's principal function for 150 years. That history was closed, and Kessler and his fellow graduates looked out on a changing world. The United States was now part of global affairs with its web of commercial and political interests. It would be the job of the navy, already advanced in its construction of modern warships, and the army to protect a bustling and productive nation both at home and wherever its interests might lead. The army was not far behind the navy; Congress also had begun to appropriate large sums for the construction of great concrete fortifications and impressive weaponry to guarantee the security of the nation's most important harbors.

Kessler knew what those changes had meant to his own branch, the artillery service. When he was a new lieutenant, the artillery was little more than infantry. In fact, it was he and a detachment of artillery soldiers acting as infantry who had captured a position near Manila during the Philippine Insurrection. The interest in coast defense brought new vigor to the artillery. Kessler was there when the soaring importance of protection from naval attack and the particular values attached to coast fortification caused the creation of the Coast Artillery Corps, a branch so specialized that it had almost nothing to do with the balance of the army. He had traveled from post to post and had witnessed the increase in fortifications like those that now flanked the entrance to Admiralty Inlet and Puget Sound.

As a coast artillery officer, Kessler had seen the good years. Now he was seeing the bad. Before the World War, Kessler told his audience, coast defense troops had made up about one-quarter of the army's strength; now they amounted to little more than one-tenth. He went on to show how the reduction had changed the defenses locally. In 1916 there had been 1,400 men to man the guns, searchlights, and fire control instruments in Puget Sound; now, ten years later, there were only 335. It was not a real military force which Kessler had to parcel out to the five forts under his charge. All he could do was take care of the equipment and buildings; there was not a sufficient number for any useful training.[2]

He did not tell the assembled members that there were other problems. Almost all the forts in the United States had lost many of their cannon during the war when armament was removed and prepared for service overseas. At the time Kessler was speaking, the ordnance inventory in Puget Sound was half what it had been before the war. Of the guns that remained in their emplacements, almost all had been installed a quarter century past. While they were the best to be had when they were mounted, naval vessels now carried cannon that far outranged them. There had been no policy to replace the guns to match or exceed the cannon that new ships might bring against them. Moreover, there was a major rift within the Coast Artillery Corps itself. Many officers felt that the idea of fixed fortifications had been so discredited during the war in Europe that their own survival depended upon an overwhelming endorsement of mobile artillery, leaving the massive permanent batteries to some antediluvian past.

Kessler continued to deliver his simple message, devoid of his own opinion except as might be expressed in the figures he gave his audience. He referred to the cost of the Puget Sound forts— ten and a quarter million dollars not including repairs, the acquisition of land, and assorted war time construction—in comparison with the number of troops now assigned to them.[3]

Needing some sort of cost estimate for his talk, Kessler had contacted the Seattle District of the Corps of Engineers. It was the members of the Corps who had designed and built the fortifications, and in its offices on Second Avenue in Seattle, it still maintained all the records of the long activity. The cost was somewhere in the collection of reports, letters, and progress statements, now yellowing with age and disinterest. The construction of fortifications had been the Corps' most important task at about the turn of the century, but it had gone on to other things, and the significance of coast batteries seemed distant when Kessler made his request. He received a broad estimate, composed by adding up the costs of the batteries, guns, carriages, quarters, and other buildings at Fort Worden. To develop the cost for the other four forts—Casey, opposite Fort Worden on Whidbey Island; Flagler, roughly in the middle of Admiralty Inlet on Marrowstone Island; Ward, far to the south on the Bainbridge Island side of Rich's Passage; and Whitman, on Goat Island near Deception Pass—the amount for Fort Worden was increased by one half. The result, added to the Fort Worden estimate, became the total for the entire system. Imprecise by the standards with which the fortifications were built, but good enough for 1926.[4]

Kessler concluded and, amidst the polite applause, returned to his seat. He had shown the place of the Puget Sound defenses within the national scheme and had described in one way how the decisions that were applied to the army as a whole had had their impact here, at the end of the line, in a relatively sequestered corner of the country. That pattern was true of the local defenses as part of a grander plan of fortification that had begun in 1885, and in some ways, long before. But Puget Sound was no backwater. Among the men of the Corps of Engineers and Coast Artillery Corps, it became known as one of the most important and demanding harbors in coast defense practice.

The lofty position of coast defense as the keystone American military policy prior to World War I eroded rapidly in the face of dwindling appropriations following the conflict, as Kessler was keenly aware. Although it might have been difficult to appreciate at the time of Kessler's talk, the fortifications over which he had command represented the ultimate in military architecture produced by the United States, both in extent and design. The story of the development of the Puget Sound forts and the high point of their use before World War I deserves special attention.

✳ ✳ ✳

Ship captains of the nineteenth century found few good harbors on the Pacific coast of North America. Then as now the land meets water in abrupt rocky headlands or in low, sandy beaches. Sheltered openings are few. With such stingy geography, it is easy to name the principal ports that attracted the mariners: San Francisco, the Columbia River, and the long, tangled network of Puget Sound near the Canadian border.

Vessels reach Puget Sound by moving eastward from the Pacific through the Strait of Juan de Fuca, the wide channel that separates the Olympic Peninsula of Washington from British Columbia's Vancouver Island. The Sound begins where the Strait ends, some sixty miles inland from the ocean shores. The transition from Strait to Sound is called Admiralty Inlet. The Inlet is a broad opening five miles wide and defined by the Quimper Peninsula on the west and Whidbey Island to the east. Below Admiralty Inlet, the deep and protected complex of Puget Sound and its contiguous waterways continues south for almost 100 miles, forming harbors and anchorages for the largest vessels.

The region was settled late. It was isolated—the only practical way to reach the country was by water—and it had been disputed territory, claimed by both the United States and Great Britain. The Treaty of 1846 had fixed the international boundary in its present position at the 49th parallel. American settlement had been encouraged before that time to add greater weight to the government's claim of national interest in the Northwest. The American population was small, and nothing really took hold until the 1850s. Then a few small villages began, their potential prosperity attached to the success of timber exports to California.

It was a grim start, with commercial centers marked by a few log and frame buildings, tide flats, and muddy wagon roads, the collection surrounded by apparently inexhaustible forests. No one could construe the scattered town sites as significant from a business point of view or as important indicators of national prestige. It was still Washington Territory, and statehood lay almost forty years in the future. But Puget Sound was an irresistible piece of the continent, and if the country was raw, it was only for a time. Confidence was the by-word, an attitude best expressed by the addition of a Chinook jargon name for the place that would become Seattle: New York Alki—New York, by and by.

Naval officers knew Puget Sound well. Britain's George Vancouver explored the area in 1792, followed by Charles Wilkes of the United States Navy in 1841. Their favorable impressions matched those of other sailors who shared the enthusiasm of the pioneer families trying to make the land prosper. Seafarers had commented frequently on the Sound's many desirable anchorages and Army officers just as frequently offered up suggestions for defending the sites.

The United States had always protected its principal harbors with fixed cannon mounted behind strong walls. Puget Sound was not a major harbor by the stretch of anyone's imagination, yet it looked like a place that might grow. Even if eluded by business, it was well placed strategically, and it seemed reasonable that the government might at some time in the future erect a naval station for the repair and maintenance of its vessels.

Beginning in the 1850s, a variety of military boards and commissions reported their recommendations for the defense of Puget Sound. The first, a joint commission of army and navy officers appointed in 1850 to review the defense of all of the Pacific Coast, thought that a canal through the slender land bridge between Indian Island and the mainland a few miles below Port Townsend would be of greater value than fortifications. In case of war, the commission suggested, ships could slip through the canal and avoid the blockading fleet that inevitably would appear off Whidbey Island. Some works could be erected at the Narrows (near Tacoma), the entrance to Hood Canal, and several other locations, but they were only to be built at what the commission defined only as "a remote period."[5]

Individual officers as well contributed their own reports that echoed the need for protection, emphasized by the British presence in nearby Canada. There was also threat from a quarter not related to coast defense: the probability of Indian attack. It was a more realistic concern than an assault by an enemy fleet, and after the Indian troubles of the 1850s, Forts Townsend and Bellingham were established in addition to the garrison at Fort Steilacoom. They were simple frontier posts and made no impact on the need for fixed defense.

General Joseph G. Totten, the army's chief engineer from 1838 to his death in 1864, formulated the most cohesive of the early plans.

Totten, one of the most formidable figures in American military engineering, examined in 1860 the "deep, quiet, and beautiful waters"[6] of Admiralty Inlet, Puget Sound, and the Strait of Juan de Fuca. He surveyed the whole of the Pacific Coast defenses, although his special fascination with the signs of British influence—the naval station at Esquimalt on Vancouver Island and the Royal Marine detachment encamped in the San Juan Islands—seemed to give extra impetus to his recommendations for Puget Sound. His plan grew from two basic suppositions: that British naval power would continue undiminished, and that there was no community or establishment inside Cape Flattery worth defending with substantial works. To deal with the first condition, Totten recommended the construction of harbors of refuge at Neah Bay and at another locality, to be selected from several possible locations. Should there be a war with Britain, merchant vessels in the Strait could flee to the harbors where shore cannon would keep marauding warships at bay. Our own navy could use the protected harbors as a base for sallies against the foe. The harbors were essential, Totten said, for without them there would be "no chance for the daring enterprise of our spirited but inferior naval force" which would compel "an enemy into incessant employment of all of his forces, into a worrying vigilance at all his ports."[7]

The need for protected harbors established, he turned to what he believed was the lesser question, the fortification of Puget Sound and its entrance at Admiralty Inlet. He saw immediately the need for forts "possessing large armaments of the heaviest guns" on the headlands of the Inlet. The bluffs of Point Wilson,

Admiralty Head, and Marrowstone Point pushed into the Inlet, but Totten doubted the ability of contemporary artillery to close the wide channel. He felt that a practical defense of Admiralty Inlet had to involve a dependable floating battery (essentially an armored platform upon which several guns could be mounted) anchored somewhere mid-channel. A floating battery was not part of the army's arsenal, which meant that an adequate defense would have to await its development, and that meant that any fortification of the headlands would occur at some distant time. Totten therefore sought a better center for the defense, and found it in the vicinity of Foulweather Bluff. That position too had its shortcomings, and would require a costly fort built on an artificial island. Farther south, he recommended nothing more elaborate than earthen batteries for the Narrows, since defenses there could protect very little.[8]

Completing his study, Totten arranged all the projects in several categories or classes and gave priorities to the individual projects in each class. In his order of importance for first class works, he ranked the harbors of refuge in the Strait of Juan de Fuca after San Francisco and the Columbia River. The Foulweather Bluff defenses appeared in seventh and ninth places. He asked that Congress appropriate a sum to initiate the work, but it did not, and Puget Sound remained undefended.[9]

Most authorities agreed with Totten that there was little to defend. For example, Captain George Elliott, member of the Board of Engineers for the Pacific Coast, examined the sparsely populated and heavily forested Northwest in 1864, and said that if anything should be protected, it should be a navy yard or station. There was no such place, and he suggested that fortification work not begin until the site of the navy yard was selected. The navy's plan for the depot lay somewhere in the dim future, but Elliott's rationale influenced many other discussions about the Puget Sound defenses.[10]

Their possible location was still undetermined in September of 1865, when the chief of engineers asked the secretary of war to reserve through presidential executive order some twenty-four potential fortification sites. The reservations were scattered from the San Juan Islands in the north to Point Defiance (on the eastern shore of the Narrows) in the south. The logic was that since the course of settlement was still fluid and undefined, it would be best to cover all possible properties that might be needed for future use.

Reviewing the list several weeks before the executive order was issued on September 2, 1866, Major General Henry Wager Halleck pronounced many of the sites useless and already covered by perfected claims. Halleck commanded the Military Division of the Pacific, the Army's major administrative unit on the west coast. An engineer with a reputation as a military scholar, he had in 1862 preceded Ulysses S. Grant as Commanding General of the Army. Halleck wrote vigorously to the adjutant general; the great number of sites was unnecessary because the essential place to be fortified was Admiralty Inlet, and that could be done handily by arming its most prominent headlands. "On these three points," he said, "all the fortifications required for the protection of Admiralty Inlet can be constructed."[11] The statement made Halleck the first to describe accurately the future form of the defenses. He was not a visionary—he also thought it improbable that a navy yard would be established in the Sound—but he was an excellent engineer who appreciated perhaps more than his fellows the improvement in cannon that had to follow the experiences of the Civil War.

The executive order set aside reservations of 640 acres each on Point Wilson, Admiralty Head, and Marrowstone Point. With the exception of the last, the lands were almost totally covered by valid private claims. As a result, the sites flanking the Inlet were military reservations in name only. Congress appropriated no monies to purchase the lands held by civilian owners. The secretary of war did not direct the improvement of the sites where there were no private claims, nor did he send troops to occupy them. They remained what they had been: forests interspersed with a few agricultural clearings.

During 1866, the Board of Engineers—a more or less permanent component of the Corps of Engineers that tended the preparation of the nation's seacoast defenses—advanced another survey of Puget Sound and the Strait of Juan de Fuca. It was the last formal examination and proposal to be made for almost the next two decades, and it contained nothing new.

There had been more than a dozen separate recommendations for Puget Sound; it had been studied, shuffled, reshuffled, and studied again. Little innovation was left. With the death of Totten only two years past and the entire Corps of Engineers heavily colored by his twenty-six years at its head, it is not surprising that the board generously endorsed Totten's examination of 1860. It found his report to be "as perfect and exhaustive as it was possible to make it." Although the range of artillery had increased since Totten had made his recommendations, the board gave little credence to a strong defense of Admiralty Inlet, preferring, as Totten had, to support harbors of refuge at Neah Bay and elsewhere.[12]

The board also restated the desirability of securing the possession of Vancouver Island, of establishing a naval depot that could be fortified, and of putting up earthworks at Point Wilson, Admiralty Head, and Marrowstone Point. The board was less sanguine than was Halleck in its hopes for the last suggestion. For Halleck, they were the focus of the defense, but for the board, they were to be occupied only to deny anchorages to the vessels that might blockade the Inlet. To be sure, the Inlet could be fortified, as could Totten's lower position near Foulweather Bluff. However, either location would require a fort in mid-channel that would cost in the neighborhood of four million dollars. In the board's view, a good wagon road from the Cowlitz River to Olympia or Seattle would do more for regional defense than any other kind of construction.[13]

As with the preceding reports and letters, this one received little note, and served only as the capstone to the early explorations of the possibilities for defending Puget Sound. It would be difficult to over emphasize the tentative nature of the recommendations of the 1850s and '60s. Only Totten's report of 1860 and the

Board of Engineers' report of 1866 had much chance of becoming part of a congressionally funded fortification program. Totten's plans for Puget Sound were included in the proposals although he would have been aware of the doubtful future of any fortification proposed for the location. His report contained a number of "ifs," all directed at the commercial development that would make the place worth defending. His report was the exercise of a talented designer handling a question of academic interest; his extensive analysis was a response to a technically provocative question that was both politically and fiscally implausible.

The lack of any substantial community, commercial center, or military base had precluded the fortification of Puget Sound. Even those who fashioned schemes for its defense were impressed with the lack of population and development. It was a "wilderness of waters," with shores so densely covered with timber that at least one officer formulating a plan elected to do his reconnaissance entirely by boat.[14] Some claimed that the abundance of timber and coal alone called for protective fortifications. Permanent works, however, were reserved for developed areas, not for natural resources.[15]

In spite of the wild appearance of Puget Sound, several elements of the early proposals took root and proved durable. Among them, Totten's harbor of refuge at Neah Bay had a surprisingly long life, being resurrected from time to time by local politicians and appearing in Corps of Engineers surveys at the turn of the century. The concept was by then much altered; the harbor was to serve ships seeking shelter from sudden storms rather than British guns.[16] The question of the long-proposed naval base continued to be of influence, as did the value of fortifying Point Wilson, Admiralty Head, and Marrowstone Point. To delay fortification until the site of the naval station could be settled was reasoned logic, since the fortifications would certainly have to protect the Navy's interest. Yet if guns of sufficient range could be mounted on those three points, the cannon would protect the naval station wherever in Puget Sound it might be located, a perception that seems to have been missed at that time. Ordnance available in

the late 1860s could barely reach mid-channel from Admiralty Head and Point Wilson; it was certain at the same time that there would be radical improvements in the capabilities of artillery weapons. After the Civil War, naval and military science shifted dramatically, and old shibboleths changed. The unsettled state of the art slowed progress toward the defense of Puget Sound and helped bring to a standstill fortification construction in all parts of the United States.

The shattering effects of the Civil War had substantial impacts on military technology. Almost as if with a single stroke, the cannon, forts, and vessels constructed before the war were obsolete, or nearly so. Stately brick and granite fortresses had fallen with surprising ease when attacked from the landward side by the new rifled cannon, used effectively during the war for the first time. The common cannon of the period was basically a tube from which could be fired spherical iron balls. The inner walls of the tube were smooth, and consequently such weapons were called "smoothbores." Rifling a cannon, that is, cutting spiraling grooves into the inner wall of the tube, made possible a number of advantages that smoothbore cannon did not have. First, the grooves made the projectiles spin, which increased both accuracy and range. And second, the projectile could be elongated into a pointed cylinder, carrying a greater weight of iron and explosive charge than a cannonball of the same diameter. These improvements made rifled cannon a far more destructive weapon than smoothbores of equal caliber, and they could be used aboard ships just as they could on land.

Equally important was a major change in the targets of seacoast forts. Naval vessels propelled by steam and protected by iron armor quickly replaced the wooden hulls and canvas sails of their forebears, doing away with dependence on wind and tide. Only a few coast batteries were erected after the Civil War and they were mean efforts compared with the handsome engineering that had preceded them. Not much else was possible. With the end of the war, military spending dropped to a level prohibiting any adequate response to the new weapons which might be used to attack the country's harbors. Several European nations began to develop coast defense programs that incorporated the recent advances in ordnance, and built large and expensive works. England, for example, invested in muzzle-loading rifles weighing up to 100 tons, and fitted their emplacements with elaborate steam-powered equipment for handling and loading the projectiles. Even if the funds had been available, the mood in the United States did not lend itself to the construction of lavish fortifications. Too many well-designed and well-built works had been lost in the Civil War, if not physically, then by the knowledge that they were now much less capable of defending themselves or defeating the vessels that might attempt to pass by them. The technical advances begun in the war had not yet run their course, and more changes were forthcoming. If the army endorsed any new technique of coast defense prematurely, it could find the nation protected by ineffectual weapons. And the army faced a Congress little inclined to fund yet another system.

Beginning in the 1870s, developments in ordnance multiplied and led to the salient features of the modern cannon: steel, breech loading, and a projectile sent on its way by powerful and efficient propellants. The improvements were equal to the introduction of rifling, and one authority called the combination "the greatest advance to be made in artillery between its invention in the fourteenth century and the appearance of nuclear projectiles in the mid-twentieth.[17]

Naval architects took similar steps, and many countries began to introduce new types of warships into their fleets. Freedom from sails and complex rigging permitted new ways of installing shipboard cannon. Armored turrets protected rifled guns of great power, and arrays of casemates and shields masked smaller weapons. Ships had become "veritable floating fortresses of iron and steel, filled with all the latest mechanical devices for devastation and massacre."[18]

The Board of Engineers continued the pendency of the coast defenses, waiting for the questions of ordnance to settle and an affirmative Congress. By 1880 the form of the new cannon was fairly certain, and the lessons of coast defense learned in the Civil War had been absorbed and were ready to be cast into new shapes. Waiting fifteen years had gained the military a great deal of knowledge, but the long hiatus from fortification construction had not been without cost. Foreign navies equipped with iron warships posed an impressive threat against which the United States had only the remnants of aging and antiquated defenses. Our own navy languished in post-Civil War torpor. In 1881 it began to provide for its first steel warships, indicating the beginning of a new era. However, it would be many years before there could be any fleet of substance.[19]

Without a navy capable of defeating foreign fleets, a national harbor defense program took on renewed importance. The last time Congress had appropriated any fortification funds was in 1875, and the need to rebuild was everywhere apparent.[20] The Chief of Engineers described in detail the inadequacy of the defenses as they existed, and their inability to withstand attack. There was no need to identify an enemy by name since almost any navy in the world could force an entry at any of the nation's harbors.

In 1885, Congress created a board to consider the manifold weaknesses. Under the direction of Secretary of War William C. Endicott, the board reviewed the harbor defenses and made specific recommendations for a new program of fortification. The report of the Endicott Board, issued the following year, prescribed an unprecedented character for the fixed defenses of the nation, and its acceptance led to the most thoroughly designed harbor defense system ever developed. The Endicott Board called for breech-loading guns and mortars of steel, placed to be invisible from the sea. Some of the rifled cannon—eight, ten, and twelve inches in caliber—were to be placed on special carriages that could retract the weapons behind a protective parapet after firing, denying a target for the approaching ships and permitting reloading in relative safety. In other locations, the board called for guns mounted in turrets, similar to those in use on contemporary battleships. The harbors would be further protected by fields of submarine mines controlled from the shore. As an additional weapon, specially constructed coast defense ships armed with heavy guns would cruise the defended waters. The details of the Endicott Board report would be changed as its implementation proceeded, but the basic concept remained unaltered throughout the useful life of the defenses.[21]

The argument made by the board members in support of their proposal was largely one of property. The wealth that lay open to capture and destruction could be measured in the billions of dollars. Unmeasured millions could be extracted from the nation with one of its principal ports held for ransom (today a notion that seems peculiarly romantic but which was strongly held at the time). The board was quick to state that it was "impossible to understand the supineness which has kept this nation quiet… allowing its floating and shore defenses to become obsolete and effete," thereby permitting so much property to be placed in jeopardy.[22] The answer to the threat of loss was the insurance provided by fortification. With the security of the nation's commercially most important harbors as its goal, the board recommended that some thirteen hundred guns be emplaced for the protection of twenty-nine locations.[23] Puget Sound was not among those selected.

The omission was hard to understand. After the Civil War, the Puget Sound area grew at an astonishing rate, and was fast on its way to becoming a shipping center of the first order. By the time the board was created, the transcontinental Northern Pacific Railroad had fixed its western terminus at a deep water port on the Sound. Its renown as a shipping point had spread, and the number of vessels passing through Admiralty Inlet in the early 1880s had increased steadily.[24] Puget Sound no longer resembled the splendid, albeit vacant, waterway found by earlier engineers. The Endicott Board noted in its report that the "country [was] growing so rapidly since the completion of the Northern Pacific Railroad that the place [could] not be described because of this growth," certainly a statement of favorable portent.[25] Members of the board reputedly visited Puget Sound, but they apparently based much of their final opinion on the demographics supplied by the Office of Naval Intelligence.[26] Those statistics, culled from material in the 1880 census, continued to picture Puget Sound as an underpopulated and isolated frontier.

Several months before Congress had ordered the Endicott Board into being, its coming was well known. No doubt hoping to sway the board in its proposals, Brigadier General Nelson A. Miles, commanding the Department of the Columbia, initiated in October of 1884 his own survey of the defensive needs of Puget Sound.[27]

For eighteen years, no one had commented on Puget Sound, a surprising contrast to the many reports issued in the 1850s and '60s. The army was committed to the frontier and the management of Indians; foreign enemies were not its chief concern during the decades that followed the Civil War. Miles' report, issued in 1885, reflected the technical advances that characterized the time and consequently gave it the appearance of a bold and advanced scheme. In reality, it was a casual and superficial document, and indicated that its authors had only a passing familiarity with current trends in coast defense and no real expertise in their application. The report proposed rifled guns and mortars on Point Wilson, Admiralty Head, and Marrowstone Point in much the same manner as smoothbore weapons had been recommended for the same locations many years before.[28] It did include the high point of contemporary armament, the armored turret, and suggested an installation for the low sandspit at Point Wilson. The idea was probably inspired by the mounting of an impressive turret at Dover, England, not long before.

In retrospect, Miles' report was not an important piece of work and was best interpreted through an understanding of Miles himself. An Indian fighter of many years standing, he languished at Vancouver Barracks. Arrogant and contrary, he was "possessed of a fantastic but consuming desire" to become president and favored opinions of coast defense that, while perhaps not sound, tended to keep his name in the public view.[29] Remote from the centers of power, his Puget Sound report may have served to maintain his presence before his superiors, demonstrating that he was working at the very forefront of military thought. But he had little affection for his own plan. In his memoirs, he gave as few lines reviewing his proposal for Puget Sound as he did to heliograph experiments at Mount Hood.[30]

Miles' report went forward to the Assistant Adjutant General. Included with it was an endorsement from the Division of the Pacific Engineer Office which inadvertently summarized the official attitude toward Puget Sound: fortifications could not begin there until the defense of San Francisco and the eastern ports was assured.[31] His report was referred to the Endicott Board's Committee on Ports to be Defended; there it lay unheeded. His argument that the constantly increasing commercial interests of Puget Sound were of national importance was to no avail.[32]

Only a few weeks had passed following the publication of the Endicott Board's report in January of 1886, when the Legislative Assembly of Washington Territory memorialized Congress to fortify Puget Sound.[33] It was the first of many such requests issued from a wide variety of organizations. Hardly a session of Congress in the late 1880s and early '90s was without a petition from a chamber of commerce, board of trade, or other group on the Sound.

Such appeals had no immediate effect, but in 1888 the Senate was interested enough to request from the secretary of war all the reports pertaining to the fortification of Puget Sound. The chief of engineers forwarded the materials, adding that the location of the naval station would in turn decide the location of the defenses, and suggesting some earthwork batteries at Seattle and the Narrows. Admiralty Inlet could be fortified more elaborately "when the commercial and property interest to defend become great," a statement revealing a stubborn unwillingness to recognize the economic growth of the area.[34]

Not everyone in a position of influence was willing to wait to see Puget Sound added to the new program. Oregon's Senator Joseph N. Dolph, chairman of the Committee on Coast Defense, tried to include Puget Sound in the 1891 Fortification Appropriation Bill. He described "wonderful changes" that had taken place, and felt that they were of sufficient sweep to include Puget Sound among the highest priority harbors of the Endicott Board. Accordingly, he offered an amendment to increase the board's estimates by five million dollars, but he was not successful.[35]

Dolph was joined by Watson Cavasso Squire, Washington territorial governor from 1884 to 1887, and the state's first senator after it was included in the union in 1889. While governor, Squire had reported the defenseless condition of Puget Sound to the secretary of the interior. As a senator, he was a member of the pivotal Committee on Appropriations.[36]

Squire understood the rapid development in Puget Sound and realized that it would be only a matter of time before it was added to the program. He wanted to deal with a more basic problem. Thousands of tons of high-carbon steel would be required for the manufacture of seacoast guns and mortars. Within the infant American industry, no firm had the plant necessary to produce the weapons. If the government relied on the manufacturing capacity of its own principal arsenal, it would take forty years to cast and furnish all the armament recommended by the Endicott Board. Lack of weapons would further delay the protection of the Sound, so Squire urged the erection of a gun factory on the Pacific coast, maybe even as near at hand as Kirkland, east of Seattle on Lake Washington. With a second facility in operation, cannon would become available at a far more rapid rate, and the defense of Puget Sound would be correspondingly advanced.[37]

Ultimately, improvements in production made the proposed factory unnecessary. Squire may have had other than patriotic inspirations behind his interests. He was associated with the Remington Arms Company, and had been its manager prior to his political career, as well as marrying the granddaughter of the company founder. The firm stood to gain from any large scale expansion of the military forces.[38]

Whatever his motives, Squire was a staunch advocate of coast defense as a national policy. One biographer attributed to him responsibility for a twenty-fold increase in congressional appropriations and authorizations for coast defense expenditures. He committed most of his energies to renewing the defenses, and agreed with Senator John Hawley of Connecticut that nothing was "more entirely acceptable to the great mass of thinking, vigorous, patriotic, real Americans in this country than to embark upon a thorough, scientific system of coast defense."[39]

Squire's fellow members of the state delegation supported his opinions. Senator John Allen declared dramatically (if not a little inaccurately) that a Pacific Gibraltar lay close by to undefended Puget Sound in the form of the British naval base at Esquimalt, a potential rendezvous for more than a score of powerful warships in the Pacific.[40] Senator John Wilson expressed his concern that any attack on the Pacific Coast would certainly concentrate on unprotected Puget Sound. "With all its water and approaches in possession of any enemy," Wilson declared, "national disaster would follow, and any squadron of American vessels in the North Pacific deprived of its supply of fuel would fall an easy prey, being unable to fight or run away."[41]

Adding to these voices was the navy, whose actions spoke louder than words when in 1891 it began the construction of a naval station in Puget Sound. The existence of a naval station or dry dock and the defenses necessary to protect it had been linked in the minds of military engineers for many years, although it was not until 1877 that naval officers began to consider the question. It was in that year that Lieutenant Ambrose Wyckoff was ordered to Puget Sound to make a hydrographic survey of the waters between Seattle and Olympia. He was three years at the task, and in that time he became convinced of the practicality as well as the desirability of a naval station in the area of Port Orchard. With the new navy growing and American interests in the Pacific expanding, and the only large dry dock being at Mare Island in San Francisco, the time seemed right to build another facility for the repair and refit of commercial and naval vessels.

Wyckoff enthused about the many attributes of the Port Orchard location, including its defensibility. He used what influence he had and met with some success in 1888 and 1890 when Congress authorized commissions to look into the question of placing a naval station or dry dock somewhere north of California. For the initial commission, the secretary of the navy laid out nine requirements to be met by the preferred site, including that it have a "favorable position with respect for the principal lines of defense." To learn what those lines of defense might be, members of the commission met with officers of the Corps of Engineers and came away with the understanding that in so far as Puget Sound was concerned, the engineers held that the waters of Admiralty Inlet were too deep to be equipped with underwater mines. It was a crucial opinion. Where mines were present, ships had to slow down to avoid them, and slow-moving vessels were easy targets for guns on shore. If there were no mines, the ships could continue past the shore batteries at speed, passing swiftly through the danger. In the words of the commission, "Puget Sound is not among those waters of which it can be affirmed that the art of the engineer has solved the problem of absolutely excluding the entrance of an armed fleet."[42]

The conclusion is remarkable on several counts. Apparently the engineers had given up planning a defense for Puget Sound because its entrance could not be mined; there was no mention of a defense based on guns and mortars alone. It seemed that the little-known inland waterway on the distant Pacific Coast made it a question of minor interest, recalling Wyckoff's own frustration when dealing with people who had the "densest ignorance" of the Sound and its special qualities.[43] And if the army could not adequately defend a naval station, then it was up to the navy, and the commission recommended that coast defense ships and torpedo boats be the primary means of protection, vessels that would themselves depend on the facilities of the station. The commission did not acknowledge that the report of the Endicott Board included its own designs for special coast defense vessels.

The commission was certain that the best place for the navy yard and dry dock was at Port Orchard. It could become "the citadel of Puget Sound," made impregnable by mines in Rich's Passage that would be flanked by batteries on shore as well as on Blake Island, near the entrance to the Passage.[44] The report of the 1890 commission further advocated the Port Orchard site and exclaimed as well about the booming economy of the region and its great commercial merit, concluding that it would be derelict in its duty if it did not emphasize the danger in allowing "such valuable interests to remain unprotected, a state of affairs which, until remedied, should continue to be a matter of the gravest concern to the whole country."[45] Congress agreed, and the next year appropriated the funds necessary to begin work on the dry dock as the first step in building the new navy yard.

Perhaps jostled out of its indifference by persistent citizens and members of Congress, perhaps shamed into action by the Navy's decisive creation of a modern and capacious dry dock and all that went with it, the Board of Engineers—now directly responsible for the implementation of the plan developed by the Endicott Board—again took up the question of a defense for Puget Sound. It visited the area in May 1894, found the region wealthy and active, and gave praise as generous as it had been faint before.

The board members prepared their plan precisely. The rationale for defense was wholly commercial and well in the mold of the Endicott Board. Four railroads now ran to Puget Sound, the board noted, and thousands of vessels passed in and out of Admiralty Inlet every year. It was an area of extensive resources. The board repeated a Congressional report which claimed that "lumber, coal, iron ore, gold, silver, lead, copper, graphite, limestone, granite and sandstone . . . exist in apparently limitless abundance."[46] Real and personal property in the counties bordering Puget Sound surpassed 160 million dollars.[47]

To protect such a substantially endowed region, the board determined to close Admiralty Inlet by fortifying its three headlands, as had been recommended so many times before, and added some new touches that may have been inspired by the Navy's idea of a defense based on armed ships. Powerful batteries on shore were to form the main line of defense, while drawing support from any warship that might happen to be in the new station at Port Orchard. A large number of torpedo boats also was considered essential to back up the fixed batteries.[48]

The board prescribed the principal armament for the trio of forts to be built at the Inlet: two twelve-inch and five ten-inch guns on non-disappearing, or barbette, carriages at Point Wilson, three twelve-inch and three ten-inch on barbette carriages at Marrowstone Point, and seven ten-inch guns on disappearing carriages at Admiralty Head. A battery of sixteen twelve-inch mortars was also named to each location. Moving south, the board passed by Foulweather Bluff, noting only that defenses there might become necessary someday, and planned several batteries for the defense of Seattle and Tacoma. Rich's Passage, the entrance to the naval station, was to be flanked by several batteries of eight-inch guns on disappearing carriages. The actual construction of ensuing years closely followed the board's proposal. The armament of the main battery at Marrowstone Point was altered to two twelve-inch guns and four ten-inch guns. The provision of special batteries for Seattle and Tacoma was discarded and the defenses of Rich's Passage reduced, but in the main, the report formed an accurate pattern for the protection of the Puget Sound.[49]

The 1894 plan turned all of Puget Sound into a harbor of refuge in much the same way that Totten had proposed a harbor of refuge at Neah Bay, and for the same reason. The local British naval presence in the 1890s was considered to be a formidable one, and while the possibility of conflict was slight, a war might find a large commercial fleet in the Sound. The forts at Admiralty Inlet would seal off friendly naval vessels and private wealth from a depredating navy, British or otherwise.

By perceiving Puget Sound as a harbor of refuge, the Board of Engineers had proposed the defense of a compact piece of geography rather than the erection of another unit in a national defense program. No mention was made of any strategic importance that Puget Sound might have. That approach was typical of the planning for Endicott works. The feeling seems to have been that if things were done properly in each harbor, the lot would all mesh together if need be, like the parts of a carefully constructed machine brought together and assembled for the first time.

The Puget Sound plan contained a number of bad ideas, but they seemed to fall away unassisted. The proposed use of vessels in the defense is a good example. The Endicott Board and the 1894 Board of Engineers pictured the navy as a partner in coast defense much as it had been during the Civil War. However, the fleets of torpedo boats and the occasional groups of naval vessels envisioned by the board as auxiliaries would never be built in the first instance or organized in the second. The tradition of coast defense was strong in the navy and not easily given up: the navy had begun the construction of special coast defense monitors in 1874 to enhance its ability to protect the nation's harbors. Despite that

investment, the new modern navy was headed in the direction of a high-seas force and not one meant to linger along the shoreline. It was only during the Spanish-American War that the fragile connection between the navy's ships and coast defense became obvious.[50]

Another proposal that got no further than the pages of the board's report was the preparation for special works for the protection of Seattle and Tacoma. Fort Lawton on Seattle's Magnolia Bluff was just not in the right spot for gun batteries, and a defensive line at the Narrows would do no good simply because there was no need of any invading vessel to go there. The value of defenses at that location was scant; they would be a "contingency depending on a contingency too remote for speculation."[51]

With the inclusion of Puget Sound in the national coast defense program, the prelude was at an end. As Totten said in his initial examination in 1860, everything was yet to be done, but at least now there was a beginning.

Construction: Simple Astonishment

"When we get ready to commence work, you people here will be simply astonished."
—*Colonel George H. Burton, 1896*

In the composing room of the *Port Townsend Leader*, a typesetter accidentally picked up the wrong letter for his headline and transformed an ordinary piece of journalism into a hopeful prophesy. The 1896 story, briefly describing the possible rejuvenation of nearby *Fort* Townsend, appeared to readers under the declaration "Port Townsend To Be Revived."[1] A printer's error to be sure, although nothing could have been a more accurate description of Port Townsend's desperate wish for something—anything—that would return it to the promising community it had been several years before.

Port Townsend was the Key City: port of entry, county seat, a busy commercial center at the head of Puget Sound, and the first good harbor inside Cape Flattery. All it needed to insure greatness was to become the terminus of a transcontinental railroad—a common dream of many Puget Sound communities in the late 1880s. Nowhere did the dream seem so real as in Port Townsend. There was terrific real estate speculation, and construction rattled along at a heady pace. Handsome brick and stone buildings bordered the town's main street, elegant homes appeared on the bluff above the docks, street cars hurried back and forth—all waiting for the railroad to give substance to the trappings of an affluent community.

When the railroad first came to Puget Sound, it got no further than Tacoma. Then it spread slowly northward to Seattle, Everett, Bellingham, and into Canada, hugging the eastern shore of the Sound. The tracks did not cross salt water, nor did any branch line reach Port Townsend. All the hopes lay with the railroad and when the town understood that it was going to be left without a rail connection, it began to die. Construction stopped, banks collapsed, people packed up and left, and what had turned so gloomy was made even blacker by the onset of a national depression in 1893.

The depression hurt all the towns on Puget Sound, yet Port Townsend seemed worse off; its hopes had been so high such a short time before. Townspeople still talked of some kind of prosperity, but as the years wore on, their thoughts became scattered and without a single certain plan. The railroad dominated; maybe they could build it themselves, or maybe there was hope for a branch line from Olympia. There were other visions: perhaps the government would regarrison Fort Townsend, perhaps there would be road construction, the nail factory was going to reopen (and then not).

Mixed among the rumors was an occasional remark about the fortifications recently added to the government program. Even after Puget Sound was included in the coast defense system in 1894, there seemed to be little real belief that anything was going to happen. The news was all of other things, perhaps the result of a desire not to place too much store in promises from afar. Slowly, as word drifted back that the forts were in fact coming, the community gradually began to understand what that could mean and soon the citizens could scarce find the words to express themselves.

"I cannot realize," said James G. Swan, one of Port Townsend's best known citizens and destined to become a distinguished ethnographer, "that what we have been looking for, and hoping for, and waiting for, lo, these many years, has at last been vouchsafed to us… [It] is the first glimmer of a dawn in the East, which will rise upon us to the perfect day."[2] A few months later he was still enthusiastic: "let energy and hope and thrift come to the foreground and jump into the bandwagon with three cheers for Port Townsend and good times."[3] His euphoria unabated, he declared on Christmas Eve, 1896: "1897 will be a new and happy year for this community, and we can take courage in the thought that at last, the clouds of adversity which have covered us with gloom so long, are rolling by, and the bright sunshine of prosperity promises to warm and cheer us all."[4]

Others joined Swan in his belief that the impending construction would be the answer to all the town woes. The fort work would fill up the vacant houses, businessmen would move into the best buildings in town and would double their profits and better. Never conservative in their estimates for the town's future, residents dreamed that Port Townsend would be the center of all military shipping connected with the forts. It would be the railhead of a line to be built out of military necessity and the home port of a fleet of powerful steam ferries that would transship the goods to Admiralty Head, Marrowstone Point, and other sites in the Sound. There seemed no end to the good that the forts could bring.[5]

Surprisingly, army officers did little to stem the fantastic hopes of the community, and even contributed to them. "There will be no more families leave Port Townsend," said a representative of the inspector general, "but you will be likely to see all come back who have left, and with them thousands of others from the east with capital to commence industries."[6] Nelson Miles stopped in the town for a few hours, suggesting broadly during his visit that more forts would be built in the Straits and consequently even more success would flow to local businesses.[7] Harry Taylor, a captain in the Corps of Engineers and assigned to the newly created office of Seattle District Engineer, did not join in the puffery. He had walked the streets of Port Townsend almost unnoticed

and carried away the simple truth that the people looked upon the work about to begin as their salvation.[8]

The initial task of building concentrated on the big gun and mortar batteries, followed by later and separately funded projects to construct emplacements for smaller cannon, fire control stations, searchlight shelters, and also improvements to the batteries that already had been built. As a result, from 1896 to 1917 there was almost always some sort of construction in progress that was associated with the fortifications. The greatest impact came with the earliest years of the work, since it was a new event and one that involved significant mobilization of men and materials; it was at that time that the officers of the Corps of Engineers and their civilian assistants came face to face with the intractable realities of heavy construction in often difficult locations coupled with the vagaries of a contract system that seemed to guarantee limited success. This chapter and those that follow detail those experiences.

The construction of all the defenses, and later their occupation by artillery units, had an impact on the towns close by, but nowhere was the effect as great and as lasting as at Port Townsend. The population of Port Townsend fluctuated with the strength of Fort Worden. Civilians and military alike from the fort traded with the Port Townsend merchants, and they were an important part of the area's income. The post was the only continuously garrisoned station in the Puget Sound defenses and served as a financial constant for some time. It was replaced only in the 1920s, when the arrival of a paper mill brought a new foundation, and one not subject to changing troop strength.

There was another and more lasting relationship between Port Townsend and the post, and that had to do with living together. The fort and the town occupied much the same piece of geography, and each had designs upon the other. Some civilians saw Fort Worden as an obstacle in the way of community growth. Similarly, artillery officers rankled at the limited acreage available to them, and they reached into the established neighborhoods to erect fire control stations and searchlights next to civilian backyards. The presence of the fort was therefore not always welcome. A planned real estate development never materialized because it was "too close to mortar batteries for peaceful enjoyment of home life."[9] As time passed, the reservation worked upon the community a positive effect. In the contest for land, Port Townsend became the winner. It grew around the boundaries of the fort and benefited from what would become an enduring relationship as Fort Worden the military post shifted to Fort Worden the park and choice open space. The town attitude mellowed to affection. For many years, it was common to hear long-time residents of Port Townsend speak pleasantly of events or people "out at the fort."

About five miles north of Admiralty Head, Coupeville shared a little in the development of Fort Casey. By town and island standards, the construction workers at the fort in 1898 earned "princely salaries," wore good clothes, and demonstrated that they were not "wholly insensitive to the charms of beauty."[10] However, the physical separation between the post and the town prevented any lasting relationship. Even when troops garrisoned Fort Casey, they looked toward Port Townsend as the center of their social life. Today, Coupeville is a busy community of retirees and summer visitors, many of whom make the trip to Fort Casey to roam through the abandoned emplacements and to wonder at them in casual curiosity.

There was no community at all close to Fort Flagler. Anyone staying there for a length of time, whether teamster or artilleryman, found the first signs of civilization at Port Townsend. Both Fort Ward and Fort Whitman were too isolated and too erratically manned to have much of a connection with local towns.

An anxious Port Townsend businessman, eager for his fortunes to take that upward swing promised with the coming of the fortifications, found little contentment in the summer and fall of 1896. Men were not at work at the defenses, land was not being cleared, goods were not being ferried to and fro, and the military was not present. The absence of the army was probably the most irksome. There would have been some satisfaction seeing officers of the Corps of Engineers moving over the ground, even if

construction itself were not underway. The relative rarity of uniformed men sporting the collar brass of the Corps was a feature of most fortification work; in this instance, however, the Corps was not quite ready to begin in Puget Sound. It remained to split off the area from the territory managed by the Corps office in Portland, and so to create the Seattle District Corps of Engineers.

That division occurred in 1896. During the first fifteen years of the Seattle District's existence, the most important years in the creation of the coast defenses, five men succeeded each other as district engineer. Three of these men—John Millis, F. A. Pope, and C. W. Kutz—had successful and conventionally rewarding careers in the military. The remaining two—Harry Taylor and Hiram Chittenden—deserve a longer look, Taylor at least because he was the first district engineer and Chittenden because of the magnitude of his work beyond the Corps.

Harry Taylor (1863-1930), in common with the others who would follow him to the Seattle office in the next few decades, was a graduate of the United States Military Academy. West Point was basically a civil engineering school and it prepared its graduates well for work in river and harbor improvements, the major peacetime task of the Corps. Fortification construction was a sometime thing, popular or not as the demands of Congress dictated. As larger appropriations for fortifications followed the acceptance of the Endicott Board report, the Chief of Engineers diverted an increasing amount of manpower to harbor defense projects. He estimated that the high point of the investment occurred in the late 1890s when a third of the Corps' officers were involved in fortification work—more than any other assignment.[11] During the same period, from three to five percent of military expenditures went to the building of the new batteries.[12]

Harry Taylor's experiences paralleled in large part what was happening to the Corps. After his graduation in 1884, he began to work at river and harbor projects around New York, trying his hand now and then at fortifications. More fortification assignments came following his transfer to the Pacific Coast and the Portland District, and yet more in 1896, when as a captain he became the first Seattle District engineer. He left the post in 1900, went to the Philippines several years later to construct defenses there, and returned to the New York area. With the arrival of World War I, Taylor served as chief engineer for the Allied Expeditionary Force, selecting and preparing plans for the posts of debarkation in France. He also constructed bases and training camps, managed railroads, procured engineering materials of every description, and organized a forestry service. He returned to the United States a brigadier general and concluded his career as the chief of engineers.[13]

Six years after Harry Taylor left his office in the Burke Building, Hiram Martin Chittenden (1858-1917) came to Puget Sound. They had graduated together from West Point—Chittenden third in the class and Taylor sixth—but that common beginning was about the only similarity in their careers. His formal studies at an end, Chittenden found himself detailed as assistant to the officer in charge of road construction in Yellowstone National Park, about as far distant from fortification or harbor improvements as one might care to imagine. His assignment to the park spurred in him an interest in the history of the American West that equaled his commitment to engineering. Before he reached Puget Sound, he had several significant publications to his credit, including *The Yellowstone National Park* (1895), *The History of Early Steamboat Navigation on the Missouri River* (1903), *The Life, Letters and Travels of Father Pierre Jean de Smet* (1905), and his best known work, *The American Fur Trade of the Far West* (1902).[14]

His term as district engineer, from 1906 to 1908, included the planning of the Lake Washington Ship Canal and Locks, his most important project in the state, and the building of roadways in Mount Rainier National Park. There was little left to do on the fortifications. The batteries begun by Taylor had been completed and supplemented by other construction under the care of Millis and Pope. Chittenden's own individual contribution to the coast defenses was the design of a self-closing metal shutter for fire control stations.[15] It was just as well that he was not more

involved; he did not get along well with the artillery, then rapidly taking over the defenses from the Corps of Engineers.

He suffered greatly from a debilitating paralysis (locomotor ataxia) which weakened him physically, but he had an unstoppable energy even after his illness confined him to his house. He carried on with the help of his assistants, who shuttled correspondence and reports from the district's downtown office to Chittenden's home and back again. Extended normal movement was exhausting; he described a three-day visit to the forts as a "trying ordeal." When his condition forced his retirement in 1910, he continued to write, and he also accepted the presidency of the Seattle Port Commission. He was one of the few modern figures for whom the epithet "Renaissance man" has real application. He was an outstanding engineer, just as he was an outstanding historian, and his seemingly unrelated dual passions influenced each other. To his history he brought the engineer's demand for exactness, and the historian's realistic imagination colored his engineering.[16]

Chittenden, Taylor, and the other district engineers did not work alone. A principal assistant, also from the Corps of Engineers, was at the district office as well and the two men constituted the extent of the military usually involved in the construction activity. On the ground at each construction site were assistant engineers, supervisors, clerks, draftsmen, and the many laborers themselves, all civilians.

The civilian engineers hired by the government to oversee the fortification work did not have a hand in designing the batteries, but they did have a great deal to do with how the work was carried out. Most fared well even if they did not achieve the prominence of Taylor or Chittenden. Eugene Ricksecker, assistant engineer at Marrowstone Point, became Tacoma's municipal engineer and had a part in the design of the first road around Mount Rainier, the state's highest prominence.[17] Few could match the devotion to fortification work shown by Samson Douglas Mason, who stayed with the Puget Sound defenses for twenty-five years, dying at Fort Worden just weeks after his retirement in 1923.[18] W. T. Preston served at both Point Wilson and Bean Point, and would become the only civilian to be appointed Seattle district engineer, occupying that position during 1918-1919. Once retired, he lived out his years in a modest apartment in Seattle. He was saved from obscurity when the district snagboat was named for him; after plying the waters of Puget Sound for many years, in 1983 the big stern wheeler was placed on shore at Anacortes as an historical exhibit. Working with Preston at Point Wilson was another civilian, Ralph H. Ober. Following his years overseeing construction at the Admiralty Inlet defenses, he left his job with the Corps and began his career as a bridge engineer, pausing to serve during World War I, and subsequently he was often referred to as Captain Ober. He joined Seattle's engineering department and had a fundamental role in improving the city's water supply. He is perhaps best remembered as a partner in the firm of Jacobs and Ober, the designers of the George Washington Memorial Bridge (also frequently called the Aurora Bridge), in Seattle. He is notable in another way; of the many who came to build the fortifications or to later serve in them, he seems to have been the first to marry a local girl.[19]

In one instance, the history of the fortifications and several generations of a family were connected. Ralph Ober was the uncle of D. W. McMorris, who was the assistant engineer at Bean Point. Years later, McMorris' son Harold, also an engineer, became an employee of the Corps and helped update the defenses of Puget Sound during World War II. He often worked from his father's drawings, tracing them and sending the revisions for approval to his brother Alfred, then commanding officer of the federalized 248th Coast Artillery Regiment.[20] From the beginning, the atmosphere about the work was professional rather than military, and there was a strong mutual respect between the civil service engineers and the officers of the Corps. When Harry Taylor left for the Philippines, he asked both R. H. Ober and D. W. McMorris to accompany him and assist in the work.[21]

Taylor arrived in Seattle in the spring of 1896 and by June he had Eugene Ricksecker surveying the reservation at Marrowstone Point. It was an arduous task. Rough brush and heavy timber covered almost the entire bluff behind the Point. Ricksecker had to cross every acre on foot, sometimes moving along fallen logs four to ten feet above the actual surface of the ground. He and his crew

Plate 2-1. The engineer party at Point Wilson, March 30, 1899. To the far left is W. T. Preston, and standing next to him is Ralph Ober; the woman is Mrs. Davis, the housekeeper, who served in the same capacity for Preston when he went to work at Bean Point. The others are unidentified. *Author's Collection*

dragged their instruments and secured measurements through the same terrain.[22] It all made for a hard time coupled with long hours. The engineering crew worked from five in the morning to eight at night. Ricksecker and the transit man stayed on usually until midnight, transferring the data they had collected to rough sketches. They were isolated—the only contact was via the steamer that brought over the lighthouse keeper's mail every Friday—and matters were made worse by a lack of good water. "The water here is simply awful," Ricksecker wrote to a friend; it was rainwater collected the previous winter in barrels, and full of "wrigglers."[23]

On a day too wet to work, Ricksecker rowed across to Admiralty Head, site of his next survey once he finished the job at Marrowstone Point. Pioneer farming had left large clearings but they were surrounded by dense brush and heavy timber, and just as difficult to get through as Marrowstone Point.[24]

Taylor used the surveys to help locate the individual batteries. He then developed plans and specifications which formed the basis of the construction itself. Once the chief of engineers approved those materials, Taylor followed a practice that had been repeated many times since Congress appropriated the first funds for the new fortification program in 1890. He publicly advertised the project for bids, with the promise of the contract award going to the firm which offered the best balance of competence with low cost.

Elsewhere, other officers were beginning to doubt the benefits of the contract system. Offering the project to private construction firms meant revealing a great deal about the nature of the fortifications to many people, despite the wish to keep such information as restricted as possible. Additionally, once the construction firm received the contract and was at work, it was reluctant to make any changes. That attitude bedeviled the Corps completely. The new system of coast defenses was without precedent and design change was, to the engineers at least, an expected and understandable event.[25] Often individual contractors proved inexperienced with the nature of the work expected of them. The result was friction between the civilians and the representatives of

the Corps, delays in completing the projects, and even the failure of the contracting firms. The problem became chronic not long after the program began. In every harbor where fortifications were underway, civilian contractors repeatedly asked for extensions of completion dates as they foundered in the face of building great concrete structures in conditions ranging from solid rock to brackish swamps and drifting sand dunes.

With several years of exasperating experience behind him, the chief of engineers finally declared the contract system for building fortifications "open to grave objections." He urged that the hired labor system be used instead since it was quicker, more easily adjusted to changing needs, and just as economical.[26] Harry Taylor had planned to use the contract system and had even traveled to the Corps' Portland, Oregon, office to discuss some of the details with the officers in charge of constructing the defenses for the Columbia River.[27] By the time the opinions of the chief of engineers had taken official form, Taylor had advertised for bid major work on Admiralty Head and Marrowstone Point. Taylor would not escape what had happened elsewhere. All of the difficulties of the contract system would be repeated in Puget Sound and the frustrating sequence of contracts between the contractor and the government would lead to the exclusive use of hired labor in all remaining work.

The construction of the large fortifications was a substantial undertaking. To name one example, building the main battery of guns at Marrowstone Point required clearing 28 acres of virgin timber, the excavation of 178,400 cubic yards of earth, and the placing of 18,800 cubic yards of concrete, a proposal which attracted bids from fourteen firms, some as far distant as Chicago.[28] The contract ultimately went to the Pacific Bridge Company of San Francisco, and that for the emplacements at Admiralty Head, a smaller project, to the Everett firm of Maney, Goerig and Rydstrom.[29]

The Pacific Bridge Company had been at work only a few weeks in the late summer of 1897 when its crews grew increasingly restive. It seemed to Eugene Ricksecker that the general air of dissatisfaction was due either to the excitement in the Alaskan gold fields or to the scale of wages.[30] The hourly wage for common labor was very low, a circumstance made worse by the contractor's insistence that the men board at its own cookhouse at a rate of more than half a week's wage. That was untenable, particularly since many of the workers were married and so had to support two households. The crews went out on strike in August. An increase in wages settled the strike and with the men back to work, Ricksecker began to notice a great many flaws. The contractor's small railway led from the dock to the construction site. Ricksecker believed it was too light for its purpose but in any case, the contractor had not followed the grade laid out for him and the line now contained a number of sharp angles, the rails veering to the right and left to avoid large trees.[31] Problems with the railway combined with those of faulty construction several months later when the material bunkers, large timber holding facilities for the storage of sand and gravel, collapsed and buried the cars and tracks beneath it.[32] When the contractors purchased a steam shovel to hasten the excavation of the rock-like soil of Marrowstone Island, the acts of a careless operator made it useless for almost a month.[33]

As the work slowly moved forward, and as the placing of concrete for the batteries began, Ricksecker fretted over the seemingly endless small delays caused by the contractor's labor difficulties—usually not enough men to do the job required—and indifferent attitude. Also, the firm balked at the rigorous inspections, and there were serious confrontations between Ricksecker and the contractor over the quality of the work.[34] Ricksecker's dissatisfaction grew. It was fueled by small episodes—the installation of a railing in a manner other than that specified—and larger ones as well, some of them potentially disastrous events, such as the reluctance of the work crew to help extinguish a fire which engulfed several steel cannon barrels on the beach.[35]

Matters were the same at Admiralty Head. The contractors could not or did not handle their plant effectively, and the partners in the firm had few skills in management. They seemed baffled by

any equipment that was even moderately sophisticated. Hoping to speed the project, they acquired a steam scraper to backfill the excavation at the gun battery, and tinkered and fussed for weeks preparing it for operation. When all at last was ready, the scraper was of almost no value because no one in the firm had any experience in using such a device. Philip Eastwick, the government's assistant engineer at the site, found himself giving the contractors elementary instructions that were simple to the point of absurdity: the obvious use of the revolution and charge counters on the concrete mixers, Eastwick advised the firm, was to count the revolutions of the mixers and the charges used. Maney, Goerig and Rydstrom had not bothered to adjust the counters, nor were they aware of their purpose. The entire plant was defective and suffered from leaking water valves, slipping belts, and failing clutches, not to mention a continuing shortage of sand and gravel. Little care had been used in putting up the falsework over the battery site, making it dangerous to work upon, a fact made apparent by the collapse of 250 feet of trestle and scaffolding.[36]

In May 1898 Taylor instructed the contractors to improve their plant so that it could function at the minimum level stipulated in the specifications. They began to enlarge it to the necessary capacity and made promises to repair all that was faulty. Eastwick, however, had heard their words before and was inclined to discount them. He did not expect any change from the firm because the source of its difficulties, he believed, was the lack of engineering ability among the principals. Arvid Rydstrom was the only one with an engineering background. Rydstrom understood what had to be done to correct the shortcomings, yet his partners' attitude hampered him. They were not interested in pursuing the contract with the exactness and detail that the government desired; instead, they seemed to expect Eastwick to provide the supervision necessary to complete the job.

Eastwick found J. J. Maney, in particular, uncaring. Maney was not especially bothered that water and cement leaked from the mixers, thus changing the nature and strength of the concrete being prepared. He was unconcerned about fire hazards, and several blazes could be traced to his disinterest. When installing the cranes for the emergency ammunition service at the main battery, he placed blocks of wood under the iron sockets at the base of the heavy cranes and disguised the poor supports with a thin layer of cement, an act discovered by Eastwick who labeled it "one of intense ignorance or malicious mischief." Maney had been so uncooperative that he appeared in Eastwick's eyes to be predisposed to make trouble. "If he were an employee instead of a principal," an angry Eastwick said, "I would have been, on many occasions, warranted under the specifications in discharging him as incompetent, insubordinate, and disobedient, and would have done so without hesitation."[37]

Both the Pacific Bridge Company and Maney, Goerig and Rydstrom had to have their contracts extended three times before the work was completed, a total of eight months in the first case and more than five in the second.[38] Although accepted by Harry Taylor, the work was "inferior in many small ways," the product of inexperienced workmen and construction firms alike.[39] Taylor found the continued excuses of the contractors tiresome—a Pacific Bridge Company employee explaining the projected loss of two working days in a single week offered that one day was a national holiday and the other would be required to give the "patriotic citizens of this camp a chance to sober up from an excess of enthusiasm."[40] Contractors' representatives claimed that the lure of the Klondike, the excitement of the Spanish-American War, and the curse of bad weather had all acted to slow their progress; Taylor's own thoughts were that each firm had failed to appreciate the magnitude of the work. Their approach had been faulted by clumsy and parsimonious methods, the execution hampered by a less than energetic and intelligent manner.[41]

The contractors were not happy either. C. F. Swigart, treasurer of the Pacific Bridge Company, wrote to Harry Taylor early in 1899 as their strained relationship was coming to a close. "It seems to me," he said, "we have endeavored to do our best at Marrowstone Point under as trying conditions of inspection and other considerations as any contractor was ever required to do

under the United States government . . . if our contract has been a constant source of annoyance and dissatisfaction to you, as you say, kindly consider what it has been for us."[42] He asked that Taylor place himself in the company's position for a moment, adding in barbed parenthesis "and I would not wish you had the bad luck to be in our place any longer than that."[43]

Maney, Goerig and Rydstrom was also displeased, although its experience on the whole was not as bitter as that of the Pacific Bridge Company. The work at Admiralty Head had been done under more favorable conditions—the contract required no extensive excavation and there was a more tractable labor force—and the firm had been able to secure a second contract for the construction of a mortar battery as well. In addition, the principals sold their plant to the government, handily liquidating an extensive inventory that would have been of limited value to them on other projects.[44]

Maney, however, had reservations and felt that the government was at least partly responsible if the inspecting engineers were not completely satisfied with what was done. He and his partners had never been given a complete set of the plans and had in fact never even seen them except for the brief glimpses allowed when they bid the project; they worked wholly from the detailed drawings issued by the assistant engineer at the site as the work progressed.[45] The many changes ordered by the designers cut severely into the profit estimated by the firm when it entered the work, and Maney ultimately sought restitution through court action. He was resentful, and while attempting an elective office at the same time his partnership was involved in the Admiralty Head defenses, he declared that "there was more sport about politics than securing a government contract for fortifications."[6]

While work was underway at Marrowstone Point and Admiralty Head, Harry Taylor was trying to secure enough land to begin construction at Point Wilson and Bean Point, on the southern end of Bainbridge Island. Many Port Townsend residents owned land at Point Wilson, and the government was ready to pay cash for what it wanted. With such a well-heeled buyer, it was tempting to hold out for a higher price than what might be expected otherwise. The local press admonished citizens to keep prices down lest the "main station"—and the providential gift of renewed prosperity—be moved to some other location.[47] Property owners at Bean Point combined to ask prices far in excess of what Taylor believed to be reasonable, an action he countered by hiring authoritative witnesses to examine the properties and then to testify to their actual value in court.[48]

Titles secured, work began at Point Wilson in September 1898, and at Bean Point in February 1900.[49] At these and the two other sites in the Sound, the Corps of Engineers now acted as its own contractor, hiring and firing men and performing its own supervision. Its dependence upon private builders was firmly past. However, it soon began to appreciate some of the difficulties had by the two private construction companies. The greatest problem for the government, just as it had been for the earlier contractors, was the maintenance of an adequate labor force.

The major industries in Puget Sound were lumbering, fishing, and farming, all of which used large amounts of unskilled labor. With jobs plentiful, the workers needed for the fortifications were hard to come by. At Admiralty Head, centered in a substantial agricultural area, men were impossible to get during threshing season, and the laborers at Point Wilson were often sailors putting in a few days' work between ships.[50] The already limited labor force also suffered from the effects of "Klondike fever" at exactly the time that work on the fortifications was getting started.

Eugene Ricksecker, in Seattle on his way to Marrowstone Point in August 1897, discovered just how hard it could be to get someone and to move him where he was needed. About to leave the city, Ricksecker found for the second time that he lacked two men because of the Alaska gold fields. He searched for replacements, and finally located the only chainman that appeared to be in town, but the man's blankets and clothes were three or four miles back in the woods on the eastern shore of Lake Washington. Ignoring the inconvenience, Ricksecker hired him and told him to rent a steam launch at government expense so that he could retrieve his belongings. When he reached the lake shore, the new employee was able to hire only a rowboat. His gear secured at last, he returned to Seattle and took the evening boat to Port Townsend, arriving there in complete darkness. He still had to find his way to Marrowstone Point; he searched the almost deserted streets and came up with a sailboat owner who was willing to make the trip for $2.50, five times the amount usually charged.[51]

Individual workers came and went rapidly, especially at Marrowstone Point. Few would stay more than several days. During the construction period, the mean number employed there at any one time was about sixty-five men per month, but the total employed was two to three times that figure.[52] By August of 1899, the work force was down to a dozen men, and the Seattle employment agencies had no more to send.[53] In one week, five men quit, unable to tolerate the rigors of hard shoveling.[54] Ricksecker marveled at the turnover. "New workmen arrive almost daily," he said, "remain a few days and then depart, giving no particular reason except a desire to 'move on.' When 20 or 30 arrive, from five to ten will leave daily until, out of a full force, there are barely enough left to operate the plant in even a small way. They are the most independent class of men I have seen, quitting work rather than be reproved or corrected."[55] It is likely that the great isolation of the place contributed to the continually shifting laborers, since it was a phenomenon that did not occur elsewhere.

Laborers, supervisors, and assistant engineers usually lived on the site. Since the engineers were the first to occupy a future fortification, they had the pick of whatever structure might befall them. The selection never amounted to much. A homestead cabin, leaning easily into the topsoil, served as engineer quarters and office at Admiralty Head, and at Marrowstone Point, Ricksecker found a similar shelter, its board walls "well preserved with smoke."[56] To make the place a little more habitable, he papered the grimy interior.[57]

Things were more formal in Port Townsend. W. T. Preston stayed in a hotel while he maintained an office in the Mount Baker Block, one of the community's most prominent office buildings. The office was convenient to the docks but remote from Point Wilson, and by late August 1898, Preston had found a better location.[58] This was the Bash house, closest to the works of all vacant homes and still not that far away from town. He rented it despite the poor water supply—a spring opposite the house provided the only drinking water—and converted a downstairs room into an office space.[59] Several years later, when Preston took over the work at Bean Point, the engineer quarters were in the shabby house of a former land owner. Soon all the reservations had new and often permanent structures assigned to the engineers connected with the construction. At Point Wilson, the quarters and offices were especially distinctive. The frame buildings stood close by the top of the tramway and were landmarks on the reservation, distinguished by their architecture and fine location. One was fitted with a concrete vault, secured by double steel doors. The vault was the only feature that survived the demolition of the two structures in 1970 and was incorporated into a landscape sculpture built on the site in 1987.

The labor force itself usually lived in tents until the contractors built a bunkhouse or made other arrangements. Heavy rains made camp life miserable for the engineer crew at Marrowstone Point. The tents were old and soggy and even bed clothes became damp from the constant wet. Eugene Ricksecker scrambled about and managed to come up with a small sheet iron stove, which at least gave the men a warm place to spend their evenings.[60] The pioneering period was never of much duration and it was not long before boarding houses appeared at the construction sites. A boarding house was most often a private business with various managers bidding for the activity. The boarding house master provided meals and a place to sleep, and a really complete boarding house offered its patrons a wide variety of goods for sale. The Maynard boarding house at Point Wilson, for example, sold overalls, socks, gloves, blankets, and many other items, including an incredible array of tobacco products.[61] Only at the Point Wilson reservation, immediately adjacent to Port Townsend, was it possible to live in town and travel to work on a daily basis. W. T. Preston encouraged the boarding house; those who lived off the reservation had a tendency not to show up on rainy or stormy days. A crew on hand in the boarding house also could be very useful in emergencies, amply demonstrated when an alarm at two in the morning turned out the entire Point Wilson boarding house in time to save a load of lumber slipping from a sinking scow, a feat impossible if all had lived in town.[62]

At any other location, the boarding house was almost a necessity and those choosing not to use it were left to their own devices. The alternatives could be dangerous and uncomfortable. At Admiralty Head, two workers erected a shack on the beach below the bluff and one was killed when a landslide buried their shelter.[63] One group of men declined the facilities at Bean Point, preferring instead to camp by a small stream. The stream disappeared, and the men were reduced to digging holes here and there in the stream bed to get any water at all.[64]

The number of workers employed varied directly with the number of batteries or other structures in progress at any one time. At the height of the construction period, from about 1898 to 1905, there were probably no more than three hundred men at work on the defenses. They were a diverse collection in many ways, mostly itinerant and with large representations of foreign born of all nationalities.

The Italians were more numerous than any other identifiable group. They would work for low wages and for that reason, the Pacific Bridge Company had hired gangs of them from Portland and San Francisco.[65] They were also willing to do what others refused. One of the most difficult and hazardous jobs in the construction of a coast defense battery was placing concrete; someone had to stay beneath the scaffolding in the form work to tamp it down so that it filled all the voids. Many men left rather than go into the "pits." With the loaded concrete cars rattling over the rails above, it was often only the Italians who stayed behind.[66] A

Plate 2-2. The construction railway at Point Wilson ran from the dock (behind the photographer in this view) and up a steep incline to the crest of the bluff. Rail cars were pulled up the incline by a steam-powered winch at the top of the grade. From there, the line branched out to several battery sites. The same method was used for the construction of the main gun battery at Bean Point on Bainbridge Island. This dramatic image, taken on April 1, 1899, so impressed the secretary of war that he included it in his annual report to Congress. *Author's collection*

Plate 2-3. At the top of the bluff, one set of tracks continued into the concrete mixer house. Here the sand, gravel, cement, and water were brought together and poured into large cubical mixers. A system of belts turned the mixers and, when the concrete was ready, it was dumped into small rail cars waiting on tracks at the base of the building. The structure was located to the right of Battery Randol. *National Archives and Records Administration*

"padrone" kept a boarding house in Port Townsend and his men formed fully a quarter of the force at Point Wilson, appearing in almost every kind of activity associated with the construction.[67]

Building fortifications was about the same as any other job on Puget Sound. The wages were low for the unskilled, but if a man had something to offer he could make a good return for his time and effort. A general laborer working for a private contractor could earn anywhere from $1.50 to $2.00 for a ten-hour day, six days a week; a laborer mounting ordnance could earn $2.18 a day. Foremen, depending on the job they supervised, could expect to receive from $2.50 to $4.00, and a teamster with his own team could also make the same.[68] How much a man made depended to

Plate 2-4. Some construction sites were remote from the centralized concrete plant built at each fort, and special accommodations had to be made. Such was the case with Battery Lee at Fort Flagler and as shown here at the site of Battery Mitchell, across Rich's Passage from what would be named Fort Ward on Bainbridge Island. Small batches of concrete were prepared at the precariously placed facility. The large retaining wall can still be seen from the Seattle-Bremerton ferry as it passes Middle Point. *National Archives and Records Administration*

some extent on how his employers worked him. The contractors at Admiralty Head and Marrowstone Point employed by the hour and the hours were not always regular. "We work them day or night," said a contractor, "as the weather, tides and other conditions over which we have no control permit; otherwise, it would be impossible to make any progress worth the work."[69]

When the government began to do its own contracting, things became a little better. The eight-hour day was in effect and the wages were higher with laborers of various stripe receiving $1.60 to $2.40 a day.[70] It wasn't all an improvement since government supervision was more strict, the overseers more watchful, and both Christmas and New Year's Day were working days if they fell within the week.[71]

At times the job could be dangerous. While few of the tasks involved any actions inherently hazardous, safety precautions were limited and as a consequence, accidents occurred as an infrequent if constant part of the construction activity. There were broken bones, dislocations, and a wide variety of cuts and bruises from an equally wide variety of causes that included horse accidents, fights, firecrackers, blasting, and bicycle collisions.

The greatest number of injuries involved the narrow-gauge railroads and the equipment used on them. An engineer leaving his shift at Point Wilson left his locomotive and went back to check the line of dump cars he had just moved under the material bunkers on the wharf. He leaned in

between two cars to examine the coupling and just at that time the cars moved together, fracturing his skull; he died six days later.[72] Most accidents were not fatal. Another locomotive engineer, this time at Admiralty Head, was jostled around a bit when his engine left the track, and a laborer at Point Wilson suffered a fractured leg when the concrete car he was riding careened off the end of the falsework at Battery Stoddard.[73] The high scaffolding was particularly treacherous. Planks laid along the crosspieces provided the

Plate 2-5. Scaffolding, as shown here for the ten-inch gun battery at Admiralty Head, supported both the concrete cars and parts of the forms. Two sets of track are visible running the length of the battery. In the left background, horse-drawn scrapers grade the bluff to increase the field of fire, and move the spoil to fill in front of the fortification. *National Archives and Records Administration*

only footing and there were no railings. Rain and fog left the walkways wet and slippery; to make matters worse, the cramped space was often shared with a locomotive or rail car. And then there was the height itself. A man fractured his hip in a fall from the falsework at the Point Wilson mortar battery after an apparent attack of vertigo.[74]

No one complained that the work was perilous or the conditions unsafe; complaints of any kind were rare. What dissatisfaction there was usually centered on pay and working conditions, but the laborers had a very limited ability to make changes. The strike at Fort Flagler was a singular exception and even then the wage increase it gained was temporary. The workers were so mixed in composition and interest that it was extremely difficult for them to unify enough to make a successful strike. They were without a union organization, and there was little hope that they would gain a sympathetic ear. In June 1899 the sand and gravel gang on the wharf at Point Wilson struck for 25 cents an hour and W. T. Preston immediately let the whole lot go.[75] In his mind it made greater sense to do that than risk raising the cost of construction, and he, like the other assistant engineers in Puget Sound, was usually deaf to pleas for higher wages.

A few men, dismissed under uncongenial circumstances, wrote to the secretary of war to seek redress. After the secretary received such a letter from a worker formerly employed at Admiralty Head, he asked for further details. The contractor in his reply described the complainant as incompetent and dismissed him as "the ultra type of a populist agitator" who kept trying to find out the plan of the fortification.[76] The hint at compromise was adequately damning and it could also be reversed. After a man got thrown off the job at Point Wilson for fighting, he wrote to the secretary of war and revealed that Thomas Irving, the English-born overseer who had fired him, was a loyal subject of Queen Victoria with "access to all the plans and secrets of the forts."[77] He didn't of course—no one employed on the works had more knowledge about the eventual form of the defenses than was necessary to do the job—but the letter indicated a desire to stir up trouble. It wasn't an uncommon

wish. When the engineers surrounded the construction sites at Point Wilson with a high board fence, it was done not in the interest of national security as might be expected, but rather to exclude discharged employees who tended to linger around and breed trouble.[78]

Of all the laborers who addressed their comments to one official or another, only one seemed to take a real interest in the way the batteries themselves were being built, and he didn't like what he saw. He wrote to Harry Taylor, explaining that he had been a carpenter at several of the fortifications. The work at those places distressed him. "I have built horse barns for farmers," he said, "that would have kicked me off the job" if the work had been done in the same manner.[79] It was all "irregular," he claimed, and little of it square. He closed with a general condemnation. "I never worked on a job where there was more lugging of lumber and I never worked for anybody that was a greater tyrant than A. W. Horn [overseer at Marrowstone Point] nor a worse bulldozer than the man that Queen Vic educated [no doubt a reference to Thomas Irving again]."[80]

He was right in that there was a lot of "lugging of lumber," and most all the work involved a good share of drudgery and tedium. It was hard labor right from the beginning. First the site of the battery had to be cleared down to the bare ground. The engineers then outlined the plan of the battery with small wooden stakes, and used more stakes to mark the land contours around it. Thousands of stakes were required for even a medium sized work; the engineer at Middle Point estimated that he would need fourteen thousand stakes just for the few batteries scheduled there.[81] Guided by the markers, laborers excavated the foundation by hand. Happily for the men wielding the shovels, the foundations were seldom more than a few feet below the surface, those of the mortar batteries being a dramatic exception. After the excavation was complete, carpenters erected wooden forms for the outside walls and interior spaces of the battery, and topped the site with a stout scaffolding strong enough to support the loaded concrete cars and the small locomotives that often pulled them.

Plate 2-6. Forms pulled away from the hardened concrete reveal the main battery at Point Wilson in October 1899. The battery awaits the finishers, who will smooth and carefully plaster all the rough surfaces. The view is to the north from the vicinity of Battery Randol; the foundation for a detached battery storeroom is at the left. *Author's collection*

There were few labor saving devices. The basic energy was either man power, horse power, or steam power. Strong backs were always needed, and almost every construction site had its neatly marshaled piles of shovels and flat-bed wheel barrows. The horse-drawn earth scraper called a flip or a skip was a common sight. Gangs of them hollowed out the deep mortar pits, back-filled completed batteries, finished roads, and smoothed out the final shape of the disturbed earth. To excavate the rock-like soil of Marrowstone Point, the Pacific Bridge Company acquired a second-hand steam shovel. After the firm completed its contract, the government bought the shovel and set it to work at Point Wilson. It became the object of great curiosity, many people walking out from Port Townsend along the beach to watch it at work.[82] The steam donkey engine, used everywhere in the region from shipyards to lumber camps, found ample use as well at the fortification sites.

The greatest application of steam power came in the form of the small narrow gauge locomotive. For any kind of heavy construction, a temporary railroad was almost a necessity. It was easy to build and the rails, rolling stock, and motive power were handy enough to be barged from one place to another. Every fort in Puget Sound had a railroad that carried the materials needed for the construction, sometimes including even the guns and carriages. The rails began at the engineer dock, where the sand and gravel bunkers were usually located, and ran to the concrete mixing house and then branched out to the individual battery sites. Local geography had to be taken into account, which meant a different approach at Admiralty Head. There the contractors rigged an aerial cableway to carry the cement, sand, and gravel from the beach up to the mixing house that was adjacent to the site of the main gun battery. From the mixers, the railroad tracks radiated outward to other construction locations.

At Marrowstone Point the land was fairly gentle and it was not difficult to get the tracks from the beach, although it did require a sizeable wharf. At Bean Point, Point Wilson, and Goat Island, high bluffs between the construction sites and the shore meant that an ordinary railroad could not be used; the grade required would be too steep for any locomotive. The problem was solved by building a tramway along the face of the bluff. A locomotive brought the loaded cars to the foot of the bluff, where they were fastened to a long cable which ran from a hoisting engine at the top. The cars were winched up the incline, switched to a side track at the crest, and then coupled to another locomotive, ready to be towed to where they were needed.

It was a simple solution and its utility at Point Wilson suggested to engineers that the inclined tramway there be made a permanent feature of the post. Special cars could be built with horizontal floors in order to compensate for the angle of the slope. The cars would be large enough to hold a wagon and team or about seventy men, and would glide from the beach to the top of the bluff in about one minute.[83] That sort of a device was beyond the responsibility of the Corps and lay more properly with the Quartermaster Department, and it had neither the money nor the interest to pursue the improvement.[84] The idea came up again several years later when the artillery troops were faced with the continuing need to move hundreds of tons of materials and supplies each year from the wharf to other destinations on the post. Ammunition for the big guns was the most difficult. It took about a half hour to move two twelve-inch shells to the top of the hill, two projectiles being the limit that could be handled by a team and wagon. Some suggested that the construction of an electric railway that ran from the wharf and through the barracks area to the gun and mortar batteries on the hill could solve the problem and make the moving of all types of goods more convenient. In 1908 there was great interest in the idea, but it too faded in the absence of available dollars.[85]

As work ended at each post, the construction railroad was taken up. Only at Goat Island did it remain, probably because there was no economical alternative. The island shores were rugged and the defenses limited to a single battery, all of which helped restrict the funds which might be spent on improving the access. Until the post was abandoned in the late summer of 1945, the construction tramway brought men and equipment from the dock to Battery Harrison, a trip of 475 feet.[86]

From the earliest years of their contact with fortification construction, artillery officers had thought that the small railroads had value beyond their original purpose. Tracks that ran behind the major defenses at each fort would make ammunition supply easy, as it would the movement of heavy objects in general. The Commander of the Artillery District of Puget Sound considered

Plate 2-7. On a rainy day in May 1900, a worker strides toward the first of the seven emplacements of the main battery at Point Wilson. In the foreground are the rails of the ammunition supply system, waiting to be installed in the battery interior. In the background, a crew completes the roadway, paving it with crushed brick from the old kiln already at Point Wilson when the military reservation was established. *Author's collection*

it a fine idea, and recommended it in a 1907 report.[87] Hiram Chittenden demurred. It was a good proposal, he said, but there were no funds. And without support from the district engineer, the chance that money would ever be available became distant.

When the rails and ties were removed, the grades were often converted to wagon roads. The rolling stock was transferred to other locations, usually the ship canal then being built in Seattle. What was too worn for further efficient use was sold. Some of the

Plate 2-8. The same view as in the preceding photo, about seventy-five years later. The sidewalk has all but disappeared as a result of the 1906 rebuilding, the same program which dramatically changed the appearance of the entire battery. The battery commander's station on top of the storehouse was added in 1912, and at the same time, an iron foot bridge was put up that connected the storehouse roof to the battery proper. *Author's collection*

locomotives had already been used elsewhere at the time they were acquired for fortification work and by the end of construction, they were completely exhausted. After the Admiralty Head contractors sold their "reconditioned" locomotive to the government, inspectors found the brass manufacturer's plate where it had been cast aside. It was dated 1872.[88] The locomotive was pressed into service despite its age and decrepitude. It was only a qualified success; it had so little power that it could not make the trip from the beach to the gun battery without stopping halfway to build up steam.[89]

At Point Wilson, the engineers had acquired another used locomotive, this one a veteran of the defunct Water Street railway in Port Townsend. It was a "steam dummy," a small locomotive clad in the facade of a street car, a disguise often used by street railway companies too impoverished to afford an electric trolley or a cable traction system. After it came to Point Wilson, the engine continued in its original costume, shuttling back and forth between the wharf and the bluff, still carrying the brightly painted coach of its former existence. It was soon carried up to the top of the incline and fitted with a cab that was more serviceable and conventional. The original street car shell lay in the brush near the engineer quarters for some time, an incongruent piece of landscaping.[90]

The locomotive continued on until 1910 when it was too far gone for repair. Assistant Engineer Preston was told to salvage what he could and then to destroy the locomotive.[91] Even a small locomotive was not easily dispatched, but Preston's successor, S. D. Mason, came up with a good solution: he buried it. He had dug a large hole near the siding at the base of the bluff, and when it was deep enough, his crew dumped in the engine and covered it over. The locomotive was gone but not forgotten, for in its passing was conceived the legend of the buried locomotive, one of the oft-told tales of Fort Worden.

There was nothing to indicate where the locomotive was hidden and its location soon passed out of memory, the fact of its burial, however, remaining strong. It became a homely treasure and the subject of barracks small talk in the 1920s and '30s.

As the story was retold, the size of the engine varied a great deal, as did its reasons for being underground in the first place. With the advent of World War II and the prominence of the 248th Coast Artillery Regiment in the Puget Sound defenses, the buried locomotive became the object of informal searches organized by members of the 248th, sometimes in platoon strength, but diligent spade work failed to turn up any evidence. The years passed, coast defense ceased to be, and Fort Worden became a state park. In 1973 park employees were removing the foundation posts of an old building when a shovel struck metal. The men dug down deeper, hoping to rid themselves of whatever it was that was slowing their progress. The more they dug, the more they found, and by day's end they had partially uncovered the fabled locomotive. It was carefully removed and placed aside as a potential aid in interpreting the history of the fort.[92]

Port Townsend provided a good vantage point for anyone casually interested in the progress of construction. While an unofficial observer could not visit the defenses in person, he could pick up news from the papers or from the laborers passing through town. After all, the community was in the center of things. The works at Point Wilson were just next door and the sounds of construction reached the outskirts of the city. Admiralty Head and Marrowstone Point were close enough to see smoke and steam rising from the engines there.[93]

It had been easy to keep an eye on Marrowstone Point—it was only a few miles south across Port Townsend Bay—and there was considerable satisfaction when the main battery on the "innocent looking bald spot on the summit of the hill" was finished. Townspeople had watched the lengthy construction episode and had to admit that "it has seemed at times as though the greatest celerity has not been in evidence."[94] There was truth in that comment and it applied as much to the other sites as to Marrowstone Point. The Corps' frustrations with private contractors slowed the building effort somewhat, and so did the lack of adequate labor.

Plate 2-9. The beginning of the construction of an emplacement, or "pit," for four mortars at Point Wilson in 1900. The teams of scrapers are completing the excavation and moving the earth to the top of the embankment where it will be used as backfill when the concrete work is completed. The ditch diggers are preparing a drainage trench that will be laid with tile to carry water away from the structure. *Author's collection*

Plate 2-10. About one month later, the excavation is filled with forms, scaffolding, and the trestles that support the narrow-gauge rail lines that run across the site. The chutes that hang below the tracks direct the concrete into the forms. The circular shapes in the foreground are special patterns used in the floor of the emplacement to provide footings for the mortar carriages. *Author's collection*

In addition, several more factors appeared now and again to bog down the work.

Weather could be troublesome. Storms prevented activity and an occasional cold spell or rare snowfall sometimes brought things to a halt. A baleful feature was the rain and what it could do to the soil. Normally the ground was firm and hard, so firm in fact at Marrowstone Point that plows could not break the surface and dynamite had only a limited effect.[95] Yet when the fall rains came, the traffic of men and horses churned the ground into a thick slurry of mud so viscous that the animals could scarcely stagger around in it.[96] At Bean Point, the spring rains in 1900 turned the battery sites into a sticky morass; work had to be suspended until things dried out.[97] That same year found Harry Taylor puzzling over how he was going to build roads. "The soil is of a peculiar character," he noted. As long as it remained dry there was no problem, "but when wet the surface becomes easily worked up into almost quick sand."[98]

Sometimes the lack of building supplies held up construction. Beams and other iron parts did not arrive as scheduled and their absence postponed the completion of the five-inch batteries at the Admiralty Inlet forts for the better part of a year.[99] When work began on the addition of three more gun emplacements to the original four at Admiralty Head, the lack of reinforcing beams brought progress to a halt for four months.[100] A New York firm held the contract to manufacture the beams, but it was slow. A shortage of steel and a railroad switchmen's strike meant that iron did not arrive until the summer of 1902, fully a year after construction began.[101]

Changes in armament meant even longer delays for the five-inch battery at Bean Point. Work began simply enough in 1900, but in April of the following year District Engineer John Millis received a note from the chief of engineers suggesting that the guns might never be furnished.[102] The weapon manufacturers could not fulfill their contract and Congress had not appropriated additional funds for replacement cannon. With no guns and with no prospect of any, the battery sat half finished. The summer months passed without word and Millis grew anxious: the battery was the only element of the Puget Sound project that was not running smoothly. Probably hoping to force a decision, he wrote to the chief of engineers and blandly suggested that the incomplete structure be used as a target for experimental firing. He said that there was plenty of room in Rich's Passage for a decent-sized naval vessel to get into a good firing position and the bluff behind the battery would provide a serviceable back stop. The resulting destruction would be no loss to the defenses, Millis added, as there seemed "to be some uncertainty" if the battery would ever be finished, and the experiment might provide the service with useful information about the quality of the fortifications.[103] The chief of engineers was surprised, although Millis got no other reaction. The battery remained unfinished that year and the next. Two winters left the exposed concrete in bad shape and the steel work rusted. Millis suggested a change in the type of gun, urging that the work be resumed as soon as possible. To delay longer would require repairing the wharf and continuing the boarding house, which would increase the cost of the battery to twice the original estimate.[104] In the end, new guns were provided and the battery completed after nearly three years of waiting.

W. T. Preston had learned to wait at Point Wilson. For months he had labored to erect a fine concrete mixing house, cement store, and sturdy sand and gravel bunkers, all in careful preparation for the efficient construction of the gun and mortar batteries under his charge. In May 1899 he put the plant into operation for the first time. With only one of the two large mixers operating, the yield was twenty cubic yards each hour; he estimated that both mixers could produce over three hundred cubic yards in a single day.[105]

To get the most out of the plant, Preston had to have a continuous supply of sand and gravel, and there he ran into problems. The sand and gravel contract had been awarded to the Pacific Bridge Company, the same firm that had given the government engineers so much grief at Marrowstone Point. It was now operating a gravel quarry at Hadlock, about seven miles south of Port

Plate 2-11. The torpedo storehouse at Middle Point, shown here under construction in August 1900. It was intended to store the equipment for the submarine mine field in Rich's Passage; the building was abandoned when the decision was made to construct a new group of mine-related buildings at Fort Ward in 1910. It survives today as part of Manchester State Park. *Bainbridge Island Historical Museum*

Townsend, with no greater ability than it had demonstrated several years before.

The Pacific Bridge Company plant broke down over and over again, and in one case, several men were injured when part of the plant collapsed.[106] No wonder that it was not able to provide the quantities that Preston needed. He had to have at least 450 cubic yards of material to keep operating at maximum, but the deliveries consistently fell short.[107] After Preston examined the company's outfit, he found it easy to understand the delays: it was a hodgepodge of second-hand equipment. Looking over the wheezing steam engines and leaking boilers, he declared that the "whole rig is a disgrace to anyone."[108] A dismayed Harry Taylor viewed the same plant and described it as "everything tied up with strings."[109]

By July, the Hadlock operation could churn out at best only 250 cubic yards in a single day, somewhat more than half of what was expected. Construction slowed accordingly. Preston looked at his own tidy, competent preparations, and despaired. "It seems a pity to have put in so good a plant as we have here and not be able to run it to its full capacity even once during the progress of the work," he wrote to Taylor's assistant in Seattle.[110] The next week, things were even worse as the scow from Hadlock arrived only partially filled. Two weeks later the gravel plant broke down again, and concreting came to a halt. "I seem fated not to be able to put in a full week's work," Preston agonized.[111] The Pacific Bridge Company promised to make repairs and to purchase new equipment. When those promises went unfulfilled, Preston gave up and brought in gravel from another source.

The sole force that could hasten construction was the Spanish-American War. There had been ample signs of the conflict long before Congress passed its war resolution in April 1898. In January of the same year, Chief of Engineers John Wilson ordered that all the guns and carriages sent to the seacoast defenses be mounted as rapidly as possible, even if it meant postponing the completion of work already under way.[112] In early April, Wilson dashed off a quick note to Harry Taylor. "The anticipated crisis seems to be upon us," he wrote, and directed

Taylor to put "every possible effort" into finishing the loading platforms of the emplacements so they could at least be used to serve the guns.[113] Taylor did what he could, adding double shifts and keeping men at work every possible hour of daylight.[114] There was no real threat of an attack in Puget Sound by a Spanish fleet; the chief of engineers wanted to be sure that any delay caused in mounting the weapons not be laid at his feet.[115]

All the district engineers had been encouraged in a confidential letter to "go ahead and show what the Corps of Engineers can do when an emergency arises for which our country is unprepared."[116] In truth, there was very little they could do. They could not add batteries to the defenses nor could they man them in the absence of artillery troops. Elsewhere in the nation, notably in those harbors where guns were ready, the artillery went to their new weaponry with great fervor. In the partially formed defenses of Puget Sound, the war meant little more than longer working hours. The first guns for Puget Sound would not arrive until the end of August, and even then the emplacements would not be ready for another nine months.[117]

Taylor did receive authority to create make-shift minefields, using what materials he might encounter on the open market. The only thing he could find that resembled a mine case was a beer barrel. He felt that 78 such kegs, each filled with 100 pounds of dynamite, could provide an adequate defense for Rich's Passage.[118] A minefield of beer barrels, however, was not what the chief of engineers had in mind when he encouraged his men to show how the Corps could perform in a crisis, and the project was quickly cancelled.[119]

With the end of the war in August 1898, work crews returned to a single shift. The brief clash had led to no immediately visible change in the nature of the defenses. Congress had made several generous appropriations during the war months and a portion funded the five-inch batteries at the Admiralty Inlet forts, although construction did not begin until long after the war was over.

As batteries were completed, they were transferred from the Corps of Engineers to the Artillery. At some point in the early existence of a gun battery, the War Department provided it with a name that identified it among all other batteries. The practice typically memorialized a variety of men who had died in battle or had given otherwise meritorious service, the rule of thumb being that batteries for larger guns (or entire forts) were named for generals and small-caliber batteries were named for lieutenants. The military reservations themselves received names in a similar manner. Thus what the engineers had built as "Rapid Fire Battery No. 1, Bean Point" and "Main Battery, Marrowstone Point" became, respectively, Battery William Warner, Fort Ward, and Batteries John Rawlins, William Wilhelm, and Paul Revere, Fort Flagler. Suggestions for the names came from several sources: the local engineer, the chief of artillery (and later the chief of coast artillery), and members of Congress. Not every name that was put forward was accepted. For example, one proposal would have called the Marrowstone Point defenses Fort Gibbon and those on Admiralty Head might have been named Fort George Wright. It is not clear how the final selection was made.[120]

It was not unusual for a battery built as a single structure, as at Fort Flagler, to be separated into smaller administrative components. The original large caliber batteries at Admiralty Inlet were visual entities of six or seven emplacements; their employment in the defenses required that they be separated into several firing batteries, each with its own name. Sometimes the christening did not match the needs of the Artillery. Originally, the name of Battery Wilhelm applied to all six emplacements of the Fort Flagler main battery, even though the battery was designed to cover three distinct water areas. Subsequent renaming of Battery Wilhelm and several other batteries in Puget Sound soon identified the fortifications in a manner best suited to the defense.[121]

The construction years reflected the currents of the time moving through the Northwest. The greatest was the Alaska Gold Rush. The appeal of sudden wealth reduced the available labor supply and, even after the forts were manned, continually drew men away from the defenses. As late as 1904 the inspector general cited the attraction of the gold fields as a major cause of desertion in the Puget Sound artillery units.[122] The year before many soldiers sought transfers to Fort Casey when a popular rumor placed gold at Admiralty Head.[123] While some of the larger towns on Puget Sound, notably Seattle, benefited from the influx of miners and fellow travelers on their way north, the pull of the Klondike had its effect on the building and early occupation of the defenses.

The construction episode had greater significance within the context of the Endicott program itself. The nineteen months of unpleasantness between the Corps and the civilian contractors made clear that what happened elsewhere in the nation's defenses was also likely to happen in Puget Sound. Only three major fortification structures of the twenty-nine built in the Sound could be attributed to the contract system. It had been a demonstrated and perhaps predictable failure that underscored the inability of construction companies to adapt to new types of work while faced with the paired hindrances of zealous inspection and fast-approaching deadlines. The Corps of Engineers was also at fault. It had developed substantial expertise in the erection of concrete fortifications, yet it had never conscientiously attempted to impart that knowledge to those firms it had hired to do the work. Because the Corps took so few steps to aid the contractors, it is easy to believe that it was pursuing a course that could have a single outcome: the Corps would act as its own contractor. With that organization holding the perfection of the defensive scheme as its goal, civilian contractors were at a distinct disadvantage in their attempts to produce a satisfactory product. The degree of satisfaction could be met only by the designers themselves.

3

Designs of General Efficiency

"[W]e seek general efficiency rather than absolute perfection in any one detail and undue attention may not be given any one function."
—*Major E. E. Winslow, 1920*

As we look today at the major structures built during the 1890s and early 1900s, we can see a family resemblance among all the components of the defense. Sometimes it is as obvious as sets of twins: most of the batteries for three-inch and six-inch guns share enough features in common that they easily fit the definition, and so do Batteries Worth and Moore at Fort Casey. There are batteries that could be siblings: Nash at Fort Ward, Benson at Fort Worden, and of course all the mortar batteries. Battery Kinzie and Battery Kingsbury are akin to that group as well. And then there are the distant cousins: Revere, Wilhelm, and Rawlins at Fort Flagler and Ash, Quarles, and Randol at Fort Worden. We have to wonder where they came from. Less apparent than the collection of exterior physical characteristics that allow us to make these comparisons, but in the end just as important, is the pattern of change to the interior design of these same structures. Their distinctive appearance, on the inside as well as the outside, is the result of the designers bringing forward what had been learned in the past and connecting those experiences to new ideas about armament and how a coastal fort should function.

The teachings of the past filtered through the inventions of a new age proved not to be the only lessons learned. The designers grappled successfully with the immediate challenges as they understood them, but they did not anticipate the rapid accretion of naval technology. They continued to improve the defenses of Puget Sound and elsewhere, unknowing that the arc of change was already set and moving in a way that would abbreviate their success.

The significant events considered by the men of the Endicott Board—four officers from the army, two from the navy, and two civilians—were those that took shape during the Civil War some thirty years before. A most notable fact was that Union naval forces had been able to push through rivers and harbors protected by Confederate shore batteries. The works at New Orleans, Island 10, Mobile Bay, and other places failed to stop the Union fleets. The failure was not due to any lack of preparation or competence on the part of the Confederacy; the battles were hotly contested.

Despite the guns, the obstructions, the mines, and oftentimes the presence of Confederate ships, the Union vessels broke past the defenses. The reason for their success lay in the single exception to the old axiom that seacoast forts were superior to ships: given enough vessels and a commander willing to accept the losses, any defense could be breached. By the 1890s, wooden ships were gone, but there were still iron men, and it took only a little imagination to picture the effects of a foreign armada commanded by a daring officer determined to establish a foothold in one of the nation's principal harbors. The only solution seemed to be to plan the new gun batteries so that they could combine in an intense close range fire of certain destruction. Thus the heavy batteries flanking Admiralty Inlet were arranged to look inward at each other across the narrow body of water between Admiralty Head and Point Wilson, a waiting vise of inescapable conclusion.

The war had taught another lesson. The projectiles fired from rifled cannon shattered and defeated the carefully constructed walls and parapets of looming masonry fortresses. On the other hand, hastily prepared earthworks often could take the kind of abuse that collapsed more formal defenses. Troops in the field could repair an earthwork with little difficulty; only lengthy and expensive labor could return a masonry fortification to useful service.

In the new fortifications of the 1890s, concrete would replace brick and stone as the preferred building material, but the engineers still relied on great banks of earth to protect the concrete batteries. Layers thirty and forty feet thick faced the modern emplacements, bringing to mind the 1867 remark of General of the Army William T. Sherman that "earth is the true parapet for resisting shot."[1] It is possible to argue that a fortification of the period is in large part an elaborate retaining wall for the earth protection to its front.

So the defenses that began to take shape in the 1890s had long roots that, in several important ways, connected to an earlier war fought with different weapons. The backward glance was a mixed blessing. The reliance on the cushioning properties of earth was sound logic, the arrangement of the batteries for short-range fire was another thing entirely, and worth a separate discussion later. The ties to the past were muted by the many changes incorporated in the new defenses. In the 1860s, a strong seacoast fort was a dense gathering of cannon mounting side by side in a formidable collection of fire power. The greater power and accuracy of modern cannon meant that fewer weapons could be distributed over a substantially larger piece of ground than had been possible before. Since the guns were separated more widely, each cannon could be mounted in its own emplacement, which gave protection to other guns nearby. The increased distance between guns and the scattering of the structures in which they were mounted, combined with a related desire to work the defenses into the existing land forms, signaled the end of the coastal fort as a discrete structure. The typical Endicott fortification was a seemingly random group of batteries and auxiliary structures occupying a geographic salient and invisible from the sea. It was not a fortress. Its military strength was related only to the armament its batteries contained.

With teachings of the Civil War well in mind and the products of recent technology close by, the Corps of Engineers set out to give substance to the new defense system. The basic unit of concern was the gun emplacement. A well-designed emplacement had to meet several criteria. It had to be strong enough to support the gun and its carriage (which in the largest seacoast weapons exceeded four hundred tons), and at the same time provide convenient access for the crew so that the weapon could be served and fired as rapidly and as easily as possible. In addition, the walls of the emplacement had to protect the men and the gun from the shells of the enemy. Finally, it had to be a storehouse for the ammunition and maintenance materials required by the gun and carriage. No longer would they also house the troops during peace time; the gun crews would be sheltered in conventional quarters at a distance from the fortifications themselves.

Not all of these needs could be perfectly met, at least not simultaneously. An emplacement could have been created that would have afforded almost complete protection, but the hypothetical end product would have been so restrictive that it would have diminished the value of the gun. The designs that actually did come about were the evidence of a constant desire for an emplacement type that maximized utility and efficiency.

Each district engineer prepared the plans for the batteries to be built under his charge. He received guidance from the chief of engineers through a series of documents called mimeographs. The mimeographs represented the preferred form of a battery for a particular gun and carriage. They were regularly updated as new carriages, equipment, or practices came into being. The mimeographs were often quite extensive; they included text and drawings which outlined the general dimensions of the batteries and the details of ventilation, drainage, and lighting. The district engineer added and took away from the information as he saw fit to meet the needs of the locality. Although it might seem that much was already done for him, he still had to determine the primary field of fire for the battery and for the individual guns, the position of the magazines within the emplacements, possibilities of concealment, the foundations, and communications.[2]

The chief of engineers directed those preparing the defenses not to rely totally on the mimeographs, and encouraged them to draw upon other sources that could contribute to the overall improvement of the battery. It was commonplace for the officers working in different harbors to share their designs.

There were numerous examples of the practice in Puget Sound. Harry Taylor reviewed plans of mortar batteries at Boston and Baltimore before he designed the battery at Fort Casey, and he sent a copy of the design for the Fort Worden main battery to the officer in charge of construction at Key West, Florida.[3] The Corps office in Seattle had on hand a set of drawings for three-inch and six-inch batteries at San Francisco. Hiram Chittenden asked the Portland district engineer for plans of a battery at the mouth of the Columbia River when Chittenden was contemplating additional works at Point Wilson.[4] As a result of the informal sharing, the process of design improvement was almost continuous, and far more rapid than would have been possible in a more tightly structured system. It gave the engineers at widely scattered sites ample opportunity to draw upon a diverse body of knowledge against which they could review their own ideas.[5]

In large part, the purpose of the emplacements as holders of cannon shaped their appearance. Nowhere was that more true than in the creation of batteries for weapons mounted on disappearing carriages. The value of a disappearing gun lay in its hidden position. If the advantage was to be made the most of, the batteries containing the guns had to contribute to the quality of concealment. No part of the emplacement could be visible from the sea. The top surface of the emplacements had to merge into ground around it, or as the engineers described it, to have a horizontal crest. But by making that idea a fundamental principle, the engineers gained new problems in ammunition storage and handling. The storage space had to go down, somewhere below the top of the emplacement where it would be out of sight. Just

how far down depended on how much concrete was necessary to protect the ammunition from the impact of shells from attacking warships.

Because of the dimensions of the disappearing carriage, the loading platform (that part of the emplacement used by the gun crew to maneuver the ammunition carts) was nine to eleven feet below the highest part of the parapet in front of the gun. As it happened, eleven feet was about the thickness of concrete required for a magazine roof to protect it from bombardment. If the magazines were on the same level as the loading platform, they would have to be capped by a substantial mass of concrete rising well above the surface of the ground. The only place for the magazines then was on a different level entirely, below the loading platform. The result was a distinctive two-story appearance. The first story, at or near ground level, contained the magazine spaces. The second story contained the gun and its surrounding loading platform.[6]

The heavy ammunition was now a good distance away from where it was needed. Six-inch projectiles were the limit of what could be shifted by ordinary man power, and at that the shells were not handy: each weighed a little over one hundred pounds. Projectiles for the remaining calibers of large seacoast cannon were of course much heavier—even an eight-inch shell weighed over three hundred pounds—and there was no way to handle such weights without recourse to mechanical equipment. Some kind of lifting device was needed to move the projectiles and powder charges from the first to the second story, yet to rely on such a mechanism would leave no adequate means to supply the gun should the device fail. The dependence on an ammunition hoist was an inescapable feature of the two-story design. One of the engineers summed up the dilemma: "[w]e can build a disappearing [gun] battery to provide concealment of the location of the gun or we can build a battery so as to provide the easiest form of ammunition service, but we cannot build a battery which will provide both of these advantages."[7]

The design of Fort Worden's main batteries (named later as Batteries Ash, Quarles, and Randol) was intended to be a solution

Plate 3-1. To be completely effective, the disappearing carriage needed an emplacement that was invisible from the sea. The engineers translated that requirement into what they called a horizontal crest, evident here in this view of Fort Casey. The top, or crest, of the battery was flat, and blended in with the gradually sloping ground to its front. Once the cannon had recoiled behind the parapet, there was no target at which an enemy could aim. *Washington State Parks and Recreation Commission*

to the difficulties associated with conveying large caliber ammunition from where it was stored to where it was needed. The battery was unique and although the innovations it incorporated proved unworkable, it demonstrated the lengths to which the emplacement builders were willing to go to avoid mechanical lifts.

All seven guns of the Worden main battery were to be mounted on non-disappearing or barbette carriages. The chief of engineers had issued no mimeographs to guide the design because too few guns of the type were to be emplaced nationwide to bother preparing a standard plan. The engineer tasked with the

Plate 3-2. To the rear of a horizontal crest battery, there was almost always a great deal to indicate its existence. The two-story appearance was a hallmark of batteries for cannon mounted on disappearing carriages. These emplacements at Fort Casey also show the results of the many efforts to keep them up to date, including enlarged loading platforms, protection above the ammunition delivery areas, and the prominent battery commanders' towers. *Author's collection*

building of such a battery in his district could either adapt a standard design for disappearing guns to fit the barbette carriages—by far the easiest and most often selected course of action—or he could strike out in an entirely new direction. It was the new direction that appealed to a lieutenant assigned to the Seattle District Engineer's office and his handling of fortification forms produced one of the most unusual batteries constructed in the United States.

M. L. Walker (he almost never used his given name, Meriwether Lewis, to whom he bore no relation) was Harry Taylor's principal assistant. In the spring of 1898 Walker had visited the defenses of San Francisco and had come away much impressed with a battery of barbette guns. The ammunition delivery system was very unusual. The hoists of most batteries opened on to a lobby, a small space in the rear and on the flank of the emplacement, but in this particular battery there was no lobby. Instead, the hoist shaft ran up through the center of the battery and opened on to the loading platform through a small doorway near an interior corner.

The result was that the hoist could deliver a single loaded ammunition truck nearer to the gun than would have been possible had the hoist been located in the rear of the emplacement. Moreover, the crew maneuvering the truck enjoyed greater protection from enemy fire because they were able to stay relatively close to the interior wall. Walker suggested to Taylor that a somewhat similar design could work at Point Wilson. With Taylor's blessing, Walker came up with a proposal which, although ostensibly based on the San Francisco battery, resembled its inspiration as little as it did any other structure in the entire coast defense system.[8] Taylor was enthusiastic about the new idea and promoted it so strongly that the battery soon became identified with him rather than his assistant.

The design, a complete departure from existing practice, contained a number of dramatic changes. Since the guns were to be mounted on barbette carriages, Taylor reasoned that a horizontal crest would not be necessary. He was thus able to bring

the magazine spaces and galleries up to a higher level than might otherwise have been possible, protecting them by a gently sloped roof of concrete that rose above the surface of the ground. Below the heavy roof, three rails attached to the ceiling of a central passageway led out from the magazine, located on the flank of each emplacement.

The rails terminated at the gun block, a large drum-shaped mass of concrete to which the gun carriage was bolted. In almost every other battery, the gun block was not a distinct visual component of the emplacement. It usually appeared as part of the working or loading platform. However, in the Point Wilson battery, the gun block stood alone and fully six feet high, which had the effect of placing the gun on what seemed like a lofty pillar. Projectiles and powder moved out along the rails and once outside, they were lowered to the base of the gun block and placed upon a small cart or truck. The cart ran on tracks circling the rear of the block. The cart would be pushed to a spot below the gun and the ammunition would then be lifted up by the crane mounted on the side of the carriage above.

Taylor forwarded some preliminary designs of the battery to Colonel Charles R. Suter, originator of the San Francisco battery that was so thought-provoking. In his accompanying letter, Taylor tried to forestall any objections to the odd configuration. He acknowledged that it was a disadvantage to depend on the carriage crane to hoist the half-ton projectile almost fifteen feet to the breech of the cannon. Better that, he said, than to rely on an ammunition lift.[9]

Suter looked over the materials and wrote to Taylor that the design was a "pretty radical departure" from what had gone before, but he liked the idea. After all, he concluded, "the abolition of lifts is of so much importance as to justify almost anything."[10] He cautioned that there might be some opposition to the special ammunition trucks which were an integral part of the emplacement. Such devices were provided ordinarily by the Ordnance Department, which protected its areas of interest with a jealous tyranny.

Plate 3-3. Later batteries for six-inch guns on disappearing carriages ignored the horizontal crest. In this recent photo of Battery Valleau (Fort Casey), a large mound of earth marks the interval between each gun emplacement, which made the battery much more visible from the water side. The horizontal crest was done away with in order to reduce the structure to a single level for easier ammunition service. Projectiles were brought from the shell room (marked by the large doorway in the traverse wall) to the low banquette encircling the raised loading platform. Men standing on the banquette handed the shells up to the loading detachment. Powder was brought through a doorway in the parapet wall. *Author's collection*

Taylor understood that there might be problems. The low-slung railway car was, in his own words, "a somewhat peculiar looking affair," with its four wheels crabbed and splayed to accommodate the radius of the track. Nevertheless, it seemed to him absurd "to think that we have got to design a $200,000 battery to fit a $50 ammunition truck because it happens to be the style designed by the Ordnance Department when another $50 truck, although differently designed, will do the work just as well or better."[11] Taylor was persuasive and the chief of engineers approved the battery for construction.

The first hint that all was not well came shortly after the battery was completed. John Millis, Taylor's successor, toured the defenses with Taylor and received an introduction to the overall scheme. Millis later recalled that he noticed several unusual features about the battery; however, he did not feel it was appropriate to comment. After all, the battery was finished and ready for troops. Millis's doubts increased during December of 1900, when he attempted to construct some of the special ammunition carts. He received bids far in excess of his own estimate, which was already some four times the fifty dollar cost mentioned by Taylor; certain parts could not be manufactured locally and had to be procured from the east coast. Millis begrudgingly accepted a price that he felt was too high and three of the special carts followed.

With the arrival of the carts, the system was complete. Millis tested it with what weights that were on hand—there was no ammunition available—and it appeared to work. He hesitated to declare the design a success since he felt that only the Artillery could give it a fair trial.

In May 1902 the first artillery units landed at Fort Worden and the reactions of the officers soon gave support to Millis' own opinions. Within several months, formal complaints began. Artillery officers commented that the height of the gun affected the rate of fire, and its location above the floor of the emplacement caused additional concern. It made work on the carriage difficult because of the limited space on the gun plug where a man might stand. There was no railing to prevent a fall, nor could there be without limiting the traverse of the carriage.

Plate 3-4. An emplacement of the original Fort Worden main battery, notable for its ingenious, but ultimately unsatisfactory, ammunition delivery system. The gun sits on a pillar high above the emplacement floor. The opening in the right traverse wall is the entrance to the ammunition storage areas, and trolley rails run from the opening to the pillar. The special cart to carry the powder and the projectile is hidden in the shadows below the trolley rails. *Author's collection*

Some also objected to the sloped concrete roof over the magazines; they feared that shells striking the surface would shower adjacent emplacements with dangerous fragments of concrete. By September, the battery was under review by the Board of Engineers.[12]

The complaints of the artillery and the suspicions of Millis were well-founded. The solution that Taylor had claimed "to be about as simple and direct a method of handling the ammunition as could be worked out" was in actual practice very limited.[13] The design did not do away with vertical lift; it merely removed it from the Corps and placed it, quite literally, in the hands of the artillerymen operating the winch on the gun carriage.

The duties of the engineers in delivering ammunition ceased when the projectile and powder reached a point where they could be transferred to a maneuverable wheeled truck for final movement to the gun. That point normally occurred in the lobby, where the projectiles were taken from the Corps-installed hoist and deposited on the trucks manned by the artillery gun crew. Because in Taylor's battery the carts replaced the trucks, and the carts were part of the emplacement, there was no formal lobby and the departure from more familiar fortification forms muddied the responsibilities. Building special ammunition trucks usurped a function of the Ordnance Department, whose representatives also criticized the design of the battery. The reaction of the artillery officers contributed to the pattern of friction between the Artillery and the Corps in Puget Sound, an unfortunate attitude of mutual misgivings which colored the early relationships between the two branches.

The question of ammunition service aside, Millis felt that the battery was the "best piece of work" of all the fortifications in the Sound, and many other officers considered it one of the best constructed on the Pacific Coast.[14] Despite such attributes, the lack of a practical method of ammunition service meant that some form of mechanical delivery had to be installed. Millis redesigned the battery, and the modification from experimental to the army's by then standard method of ammunition delivery began in 1904.[15]

The changes altered the appearance of the battery so greatly that there was little similarity between the original design and that which would serve the guns for the majority of the useful life of Fort Worden. The entire surface of each emplacement to the rear of the gun block received a fill of beach sand topped with a layer of concrete sufficient to raise the floor to the level of the gun block. To make the emplacements somewhat larger, the raised surface covered areas in the rear formerly used for sidewalks; steps led up to the new emplacement floor.

The most striking changes came with the addition of the ammunition hoist. A large rectangular block of concrete was removed from the flank of each emplacement to form a lobby, and from the lobby, a single hoist shaft was cut through the concrete mass to the magazine space below. An unusually heavy roof, supported by three massive pillars, covered the lobby. (The artillery officers were not quite satisfied; they had hoped for modifications that would have permitted direct loading from the ammunition truck into the breech of the gun.[16]) The alterations, completed in 1906, fundamentally modified its initial form and left the battery with only a handful of indicators to its former appearance.[17] However, even in its reworked form, the Fort Worden main battery was a singular structure that stood apart from others built during the same period.

The attempt of Walker and Taylor to incorporate horizontal ammunition service into a vertically organized battery was perhaps a predictable failure. Horizontal ammunition service was possible only when the ammunition store and the gun were on the same level or nearly so, and when the earth and concrete cover needed to protect the magazines would not reveal the location of the battery. Only emplacements for the stubby seacoast mortar could meet those conditions.

The efficiency of mortar battery ammunition service made it easy to appreciate the desire of the engineers to create a similar service for other coast batteries. There was no vertical lift except that necessary to raise the half-ton projectile from the magazine floor to the ammunition cart. The emplacement of the cannon and the magazine were adjacent to each other and at the same level; the trip to the breech of the mortar was short indeed.

Horizontal ammunition service was possible in mortar batteries because the mortars were an entirely different sort of weapon. They fired their projectiles upwards at a steep angle. It did not matter what sort of obstructions lay between a mortar battery and its target; the trajectory of the projectile was so high that it could clear almost any natural or manmade feature. The ammunition stores could be placed close to the mortars because the interior rooms could be protected generously without compromising the other design needs. In Puget Sound, the mortar batteries were built into the reverse slopes of hills, which hid them and utilized the existing topography as a shield for the magazines.

At first glance, the mortar batteries at Forts Casey, Flagler, and Worden are disappointing to a visitor who has seen the complex forms of the batteries for guns on disappearing and barbette carriages. Instead of great blocks of concrete, deep hoist shafts, uncountable niches, and a labyrinth of rooms, each mortar emplacement seems little more than a paved box, set deep into the ground and open to the rear. The walls lining the box (called a pit in the parlance of the coast artillery) are mostly unbroken by openings save for a large doorway on each flank. The doorways give way to a single interior corridor that follows the perimeter of the emplacement. Most noteworthy perhaps is the depth of the pits at Fort Casey and at Fort Worden; the emplacement floors are almost forty feet below the crest, an indication of the massive earth protection.

The total impression is one of casual simplicity. The apparent lack of sophistication is deceptive, for the mortar battery was one of the most imaginative features of the defense program. The emplacements affected how the cannon placed in them were to be used, and thus formed a rudimentary weapons system.

The mortar battery was an old idea in coast defense, dating from 1809 in this country; the mortar was itself far older and originated not long after the introduction of gunpowder.[18] However, mortars were never more than auxiliary weapons until

the American Civil War. Then they became key elements of land warfare, and proved their worth by destroying earthworks that were impervious to the bombardments of guns. Their value in coast defense still seemed questionable. To be sure, almost every seacoast fort of the period had its complement of mortars, yet they were often ignored in favor of other types of cannon. However, during the Civil War, the nature of warships changed in such a fundamental way that it made the mortar a singularly necessary piece of ordnance. With the introduction of iron armor, it became increasingly difficult to defeat naval vessels. There remained a single unprotected opportunity. The decks of ships were not covered with armor, nor could they be and still remain seaworthy. Mortar shells could rain down to pierce the decks of the strongest warship.

At least in theory. The mortars of the 1860s were so wayward that there was doubt that they could ever actually hit anything as small as a ship. The potential still seemed great, and after the war they were the subject of continuing discussions about their utility as seacoast artillery. The United States could have a formidable weapon if the inherent inaccuracy could be resolved.

The most workable idea came from an officer of the Corps of Engineers who had commanded a battery of mortars at the siege of Petersburg. General Henry L. Abbot proposed that mortars not be treated like guns. Instead of being aimed and fired one at a time, they should be grouped together and fired at once, like an immense shotgun. By firing many shots instead of just one, the chances of one of the mortar shells striking a ship were considerably improved. As an illustration of what the group of mortars might look like, Abbot built a demonstration battery in the early 1880s, providing the direct ancestor of all the mortar batteries built by the United States in the subsequent decades. Abbot's battery consisted of four deep pits, each pit placed at the corner of an imaginary rectangle. Each pit held four mortars. Underground galleries connected the pits and provided magazines and storerooms.[19]

The size and shape of the rectangle was the key to success which Abbot claimed for the battery. He conceived its proportions to adjust for the erratic fire of the cannon and to ensure a more regular pattern of dispersion around the target. The group of four pits, four mortars to a pit, constituted a single weapon. What the mortars could not accomplish as individuals, they could do in numbers by being combined into a specialized battery that compensated for their shortcomings.

When it came time to install the mortars in accordance with the wishes of the Endicott Board, it was Abbot's design that the Corps of Engineers selected. Problems appeared, however, after only a few of the new mortar batteries had been built. The pits themselves were too small, and it was difficult for the crews to load the weapons in the cramped spaces. More significant, the effect of all the mortars going off at once created a partial vacuum; the sides of the pit were so high that air could not freely flow back in once it had been blown out by the force of the discharge. The blast effect tore iron doors from their hinges, burst electric lights, tumbled earth into the pits, and worked general havoc on the battery and its occupants. It was not long before the enclosed and tightly ordered pits were superseded in design by a linear arrangement of pits with the entire rear of each pit open and unobstructed.[20]

That change was due in large part to the work of Lieutenant John T. Honeycutt. He reasoned that the modern mortars of the 1890s were far more accurate than those available when Abbot had prepared his first designs. Honeycutt felt that Abbot's battery, literally adhered to, would work to the detriment of the new, more efficient mortars. He suggested that if the pits were placed side by side in a row with one side open to the rear, the damaging blast effect would be much reduced. The combined fire of the four pits in a row would be just as

accurate if not more so as the four pits in a rectangle and would be far safer for the crew. He also noted that there had been no justification of the typical battery by experimental firing or mathematics, and Honeycutt's arguments were well buttressed with discussions of mathematical probabilities.

A board of officers convened to examine the new ideas. The members concluded that the virtue of the existing design lay in its ability to compensate in a limited way for errors made in locating the target. The method Honeycutt advanced would concentrate the fall of the projectiles on the ship, which was desirable as long as the location of the ship was accurately known. What the

Plate 3-5. The mortar battery at Fort Casey featured emplacements of about the same small size as the earliest mortar batteries built under the Endicott program. The restricted dimensions made it difficult to load all four mortars easily or quickly. Artillery officers heavily criticized the battery for its shortcomings. *Author's collection*

Plate 3-6. Two men pose on the steps leading to the headquarters of the Fort Casey mortar battery in January 1900. The headquarters included small chambers for the battery commander, an orderly, a chart room, and a relocating or plotting room. The same grouping was included at the Fort Worden battery, albeit at ground level, and was done away with entirely in the design of the mortar battery at Fort Flagler. *Author's collection*

board did not mention was Honeycutt's belief that the only thing that stood between his design and the ultimate perfection of the mortar battery was an improved rangefinder. The officers of the board knew, as did Honeycutt, that substantial advances had been made in recent years in optical position finding equipment, and improvements were continuing. To insist upon a particular configuration for a battery because it worked well with a soon-to-be-obsolete piece of apparatus was faulty thinking. In quiet recognition of superior design, the plan of the mortar batteries changed in the mid-1890s from four pits arranged in a rectangle to four pits laid out in a line.[21]

The mortar batteries in Puget Sound, like the majority of the mortar batteries built, were of the linear variety. Changes and improvements continued even after the arrangement of the pits was settled. The batteries at Forts Casey, Worden, and Flagler were designed over a period of years, and consequently the battery at each post is a moment in the episodic growth of the mortar battery as employed in the nation's defenses. They are significant, too, for their portrayal of regional adaptations.

The battery at Fort Casey numbered among the first constructions of the "four in a row" type. In some ways, the design harkened back to the appearances and flaws of the pits used in the old rectangular plan. The pits at Fort Casey were small, and in fact were no wider than those in Abbot's design, a tight forty feet that had to accommodate two mortars abreast. Each mortar sat on a circular platform eighteen feet in diameter; there was precious little room to spare between the four cannon grouped in the forward end of the pit.[22] The pits were longer, but added length did not make it much easier to load the weapons. Above the pit, the sloping sides of the parapet and traverse were steep and high.

The artillery officers assigned to Fort Casey were quick to note the drawbacks. In 1900, the coast defense commander reported that the size of the pits was "so limited as to greatly impede and inconvenience the maneuvering of the detachments and pieces" and believed that the cannon could be "worked only at a disadvantage." The parapets he found to be excessive, and the water

and mud from the fall rains poured off the slopes and into the emplacements, and almost onto the mortars themselves.[23]

Adverse comments continued until, in 1903, a representative of the chief of engineers asked if any of the cannon actually had been fired and what ill effects had occurred. There had been some recent practice and the results were not good, or at least they were not what the engineers wanted to hear. The discharge of one mortar blasted out a layer of concrete from the parapet and scattered clouds of grass, dirt, and dust about the pit; another blast threatened to set fire to the grass around the battery.[24] The next year, the inspector general declared that the entire battery would be of very limited value under service conditions, and might not even be safe.[25] Nothing could be done. The battery had to remain as it was or be torn apart and built again from the ground up, which was not a realistic alternative.

Two years after Harry Taylor and his assistants designed the Fort Casey mortar battery, they began work at Fort Worden, and the battery there had none of the shortcomings of their earlier effort.[26] Taylor followed the new general specifications for mortar batteries that reflected the Artillery's desire for more generously dimensioned emplacements. Thus the 1899 design for the Worden battery featured pits more than half again as great as those at Fort Casey. They were still deeply placed, although the parapet and traverses were sloped back and away from the muzzles of the cannon. The large pits were amenable to the Artillery, and spaciousness became the pattern for all mortar batteries to follow. The Fort Flagler battery was built several years after that at Fort Worden and it demonstrated few differences, the most apparent being the reduction in the amount of earth cover, the result of being built into a shallower elevation than the batteries at Casey and Worden.

The steadily accumulating experience of the constructing officers meant that change was a common quality in all the fortifications. The improvements manifested themselves in two principal ways. One was the periodic revision of the standard plans used to guide

Plate 3-7. While one mortar points upward in its firing position, a detachment of the 108th Company goes through a loading drill "by the numbers" at Battery Brannan. The emplacements of the mortar batteries at Fort Worden and Fort Flagler were much larger than those at Fort Casey, and the greater dimensions made the cannon much more accessible. *Author's collection*

the design of wholly new batteries; the growing size of the mortar pits was in some ways an example of such incremental change. The impact of changes incorporated in any one current plan might be slight, although the net effect over a period of time could be substantial: the standard 1894 design for a battery of twelve-inch disappearing guns bears little resemblance to Battery Kinzie, built about fifteen years later for the same type of armament. The other method of introducing change was through the enhancement of existing batteries by modifications large and small that helped keep older works on a par with newer installations. Yet the most vexing problem had nothing to do with new construction or modifying what already had been built. It had to do with the nature of concrete as a building material.

Despite the impenetrable appearance of concrete, water travels quite easily through it. The Puget Sound batteries had a considerable advantage in that they were made with an excellent grade of imported European cement (as was almost all concrete work in the area prior to the development of a local Portland cement industry), and the problem was considerably reduced, but not eliminated.[27] Because concrete is a poor conductor of heat, any exposed surface expands and contracts at a far greater rate than the dense mass below it. It was not long before the batteries were veined with hairline cracks. Rainwater entered the cracks, percolated through the concrete, and eventually dripped steadily into the interior rooms and galleries, leaving mineral deposits on the ceiling, walls, and floors.

The seepage was unsightly. It also promoted rust on iron and steel equipment, led to the deterioration of electrical machinery, and could even eat through the lead shielding on power cables. Moreover, water in the cracks froze and expanded during winter which enlarged the cracks, admitting greater amounts of water. Cracks and seepage could occur with surprising speed. By May of 1899, immediately after the completion of the gun battery at Fort Casey and even before the guns were mounted, the assistant engineer had detected a crack in the loading platform and evidence of seepage on the interior.[28]

The most effective way to prevent seepage was to stop the water from entering in the first place. Waterproofing was a fairly easy matter in those instances where the treated surface was to be covered with earth. Heavy applications of tar faced with drain tile directed water away from the concrete.[29] The first batteries in Puget Sound did not have any sort of subsurface waterproofing, but it was a simple matter to remove the earth and apply the protection. Thus in 1903 a portion of the parapet over the Fort Casey mortar battery was removed and several coats of tar were applied.[30] At Fort Flagler, the front wall of the main gun battery was excavated and the surface treated with a layer of tarred burlap, a layer of common brick, and a final veneer of hollow tile.[31]

Still, only a relatively small area of any battery could benefit from subsurface protection; major portions were open to the weather and all the forms of moisture it could bring. Waterproofing preparations applied to those exposed areas yielded only a limited benefit. A favorite approach called for cutting out cracks and filling the depression with hot tar. The tar usually blocked the water for a time, but eventually it lost its plasticity and water again seeped through. Some authorities recommended boiled linseed oil. It seemed to work, yet it had an unpredictable and sometimes adverse effect on the concrete.[32] A few commercial products found their way into use, but not many. A fabric was applied to the parapets of the main batteries at Fort Worden and Fort Flagler, and then treated with a preparation called Ebonal.[33] Like all the other preparations applied to the surface, it was only a limited success and not used again.

The seepage could render the interiors almost unusable. The condition was pronounced at the Fort Casey main battery, and there the assistant engineer noticed that after every practice with the guns, there were more leaks in the emplacement. He did not think that the leaks had been caused by new external cracks, and believed instead that the shock of discharge had some sort of racking effect on the internal mass of concrete that opened up additional avenues for travel.[34]

To sidestep the mysterious defect, District Engineer John Millis suggested an alteration of the battery itself. Millis proposed that the magazines be lined with cedar and fitted with an interior roof of copper, a room within a room which would divert the seep water from the storage space. The Fort Casey battery was only one of many among the nation's defenses with wet interiors, and the chief of engineers endorsed Millis' idea by way of experiment. Millis fitted up two rooms (the magazines in the first emplacement of Battery Worth and the second emplacement in Battery Moore), and kept an accurate account of the cost to see if the method might be applicable on a larger scale. The lining was completed in 1906 but the price—considered high at a little more than $2,000—was expected to increase substantially in areas where cedar was not easily available.[35] The modifications also reduced the amount of available space, which was small to begin with, and the idea was dropped.[36]

Modifications of any kind, whether the addition of special linings or the application of surface waterproofing treatments, could do little to relieve the dampness of existing batteries. More satisfactory was new construction that incorporated design features specifically intended to address the problem. Not only did the new designs remedy the presence of seep water, but they controlled condensation, another nagging source of moisture.

The engineers understood that better ventilation would help keep things dry, but there was little success in that direction until 1903 when the standard design for ten- and twelve-inch gun batteries included a system of air passages surrounding all interior spaces.[37] Not only was each emplacement ventilated by the narrow hall-like openings, but the passages were all connected and served the entire battery. Outside air was brought in through doorways in the rear of the battery and through openings in the parapet wall surrounding the loading platform. Battery Benson and Battery Kinzie, among the very last batteries built for ten- and twelve-inch disappearing guns, incorporated the air circulation system.

The two batteries exhibited another important feature. Within the large chamber designated for the magazine, a sort of house was built with thin walls of concrete or porous brick, and covered with an asphalt roof. Air could flow freely around the walls, and the roof stopped any seep water that had percolated through the concrete mass above.[38] It required some manipulation to operate the system most efficiently. The doors to the air spaces were to be opened only when the outside air was warmer than the interior of the battery. When the reverse was true, the doors were closed, and if at all possible, the doors to the rest of the battery were shut as well. A further benefit came when independent electric power plants became part of large caliber gun batteries. The gasoline engines were water cooled and produced so much heat that some provision had to be made for directing it to the outside of the battery. It took only a small change in design to locate the radiators along one wall of an anteroom at the end of the ventilation passageway. By arranging the doors of the anteroom, the hot air from the radiators could be directed through the air passages to warm the interior. Battery Kinzie is a good example of the practice.

As indicated by the introduction of the air passageways, it was in the interior of seacoast batteries rather than the exterior that there was the greatest amount of change, although it was not always easily perceived.[39] The most telling demonstration came with the arrangement of the magazine—composed more specifically of the powder magazine, the shell room, and the shot gallery—because how these three spaces were placed had much to do with how quickly ammunition could be brought to the lifts, and therefore how quickly the guns could be served. In Puget Sound there are good, bad, and indifferent manipulations of the magazine components.

Certainly the large caliber disappearing gun batteries at Fort Casey and Fort Ward must rank among the most poorly arranged. The first emplacements at Fort Casey (Battery William Worth and emplacements one and two of Battery James Moore) were built several years before the balance of the main battery. In those early emplacements, the powder magazine and the shell room were directly opposite the lifts, which was the best place for them. However, the shot gallery, the space for the storage of the heavier

Plate 3-8. An emplacement of Battery James Moore, Fort Casey, seen through the parados adjacent to Battery Valleau, about 1910. A parados was a structure used to protect batteries from projectiles that might be directed at some unguarded part of the defenses. The parados in the photo was built to prevent a shot, fired from a vessel in Admiralty Bay, from passing behind Battery Valleau and striking the rear of Battery Kingsbury. Fort Worden and Fort Casey each had a large concrete parados separating neighboring batteries. An earthen parados was more common and often employed with power plants, searchlight positions, and base end stations. *Washington State Parks and Recreation Commission*

armor piercing projectiles—the kind most likely to be used in an action—was physically isolated. There were two right angle turns between the gallery and the lifts, and to make matters more difficult, the gallery floor was some four feet higher than the floor of the rest of the magazine.

The intent had been to move the projectiles from the gallery to the lift by ammunition truck. The truck was about four feet tall, and since the gallery floor was at the same level, little work would be needed to move the projectile to the truck. Faster ammunition service was the goal, but the elevated shot gallery was too awkward to be of much help. The difference in elevation was in truth a barrier, and it was not made a part of Battery Nash, the later disappearing gun battery at Fort Ward.

Battery Nash had its own problems. Its magazines bore the mark of the ammunition service developed originally for the Fort Worden main battery in that both shot and shell were stored along the sides of a long corridor. The plan made sense in the Fort Worden battery since the projectile corridor was on the same level as the rest of the emplacement. The Fort Ward battery, however, was a two-story design. The magazine was on the first story and the gun on the second, and thus no direct horizontal connection between the two was possible. The only way for the ammunition to reach the level of the gun was through a lift. There was not enough space to install a lift at the end of the corridor so it had to be tucked in along one wall. To get the projectiles from their storage place to the lift area required that the ammunition crew maneuver the projectiles, suspended from the ceiling trolley, through a 180-degree turn of about six feet in diameter within the confines of a very small bay at the end of the corridor. Not only was it clumsy, it was inefficient and dangerous.

Such follies were the exception rather than the rule. The magazine plan in the original Fort Worden main battery was brilliantly developed despite the shortcomings of other elements of the design. More successful and more conventional were the magazines in the main battery at Fort Flagler. They bore a strong resemblance to the Fort Casey pattern, but with important differences. First, the shot gallery was nearer the lifts and not displaced to the rear, which made it possible to move projectiles from the gallery to the lifts by overhead trolley. Additionally, since the move could be made by trolley, there was no need of an elevated floor for the convenience of the ammunition truck.

The main batteries at Forts Flagler and Casey were design contemporaries, and it may be worthwhile to reflect on the marked differences in the arrangement of their respective magazines. The Fort Flagler plan was superior because of the advantage afforded by the placement of the storage battery rooms. Lead-acid storage batteries provided electrical power for motors, lights, and firing circuits, and most large caliber gun batteries had one or more rooms designated for them. In the main battery at Fort Casey, the storage battery room was a narrow L-shaped chamber fitted between the shot gallery and the gun block. Its position, nearer to the lifts, would have been much better suited for the shot gallery. However, if placed in the position occupied by the shot gallery, emergency ammunition service via the outside cranes would be hampered. At Fort Flagler, the designers escaped the same difficulty by placing a single storage battery room centrally in the traverse between the paired emplacements, rather than providing a separate room for each emplacement. It is likely that it was this aspect of the plan, as well as the ability to move men and munitions easily from one emplacement to the next, that gave it the reputation of being the "finest arranged on the coast," the oddity of having the entire first story below grade and accessible only by long ladders drawing no comment.[40]

The trend indicated in the Fort Flagler battery was an improved orientation of the powder magazine and projectile storage areas in relation to the ammunition lift. The idea matured somewhat in 1901 with the construction of Battery Kingsbury at Fort Casey. The ammunition hoists were placed in a large room between the emplacements rather than fitted into an alcove. The shell room was not a specific chamber; instead it was a designated space between the two hoists. The total impression was that the magazines and the hoist were now less formally defined spatially and were instead organized functionally. It was a more compact plan which gave the ammunition crew greater freedom of movement. Battery Kinzie and Battery Benson at Fort Worden represented the final variation of magazine arrangements in two-story emplacements. In both batteries, there was a large and separate hoist room; at Battery Kinzie it was the largest single space in the entire structure.

By the time those batteries were constructed, the design of the projectiles had changed and it was no longer necessary to provide a special room for shot storage, and with one variable out of the picture, it was far easier to manipulate the shell room and powder magazine around the hoist room. They were placed directly behind the hoist room and opened on to it through large doors. Travel from the storage area to the hoists was short and direct, and there was no need for the paths of the powder handlers and the projectile crew to cross. It was a beneficial combination gained through an improved understanding of what would work best for those serving the guns and made possible by a physically larger space in which those insights could be applied.

The changes in the interior of major caliber gun batteries emphasized a simpler and easier way to get ammunition to the gun. The revisions of the interior spaces aided that goal, but it was the perfection of the mechanical ammunition hoist and its subsequent installation in many of the Puget Sound batteries that made the biggest difference in efficiency.

The first gun batteries at Admiralty Head and Marrowstone Point incorporated a type of elevator called a balanced platform hoist to move the projectiles and powder from the magazine on the first floor to the gun on the second level. The balanced platform hoist was a simple idea. It had a pair of moveable platforms, each large enough to hold a single ammunition truck, installed in parallel vertical shafts. The platforms were connected to each other by wire ropes through the top of each hoist and running to a winch between the two shafts at their base. One platform was positioned at the bottom of its shaft at the magazine and the other was at the top of the shaft at the ammunition lobby, an alcove or area to one side of the loading platform. Set into motion by the winch, one platform rose, carrying a truck loaded with a single projectile and powder charge, while the other lift descended, carrying down an empty truck. When the lift with the full truck reached the level of the loading platform, the waiting crew pulled the truck from the hoist and ran in an empty one. When the empty truck reached the magazine level, the crew there removed it from the hoist and loaded it for the next trip up. Each emplacement had two lifts with the promise of a fairly rapid

ammunition service that provided the artillerymen a complete ammunition truck ready to move to the gun.[41]

The cycle could be repeated as often as necessary, the two trucks on the lift balancing each other, the ammunition providing the only weight actually lifted. It appeared to be the perfect system, and the Board of Engineers adopted it as standard in the mid-summer of 1896, just as the fortification program was beginning to hit its stride. As a result, the hoist was installed in large number of batteries, typically in sets of two. At first all worked as planned but it was not long before the scheme revealed serious limitations.

Each platform of the hoist rode along wooden guide rails to prevent it from twisting in the shaft. When the weather was damp, the wood swelled and the platform would jam, bringing ammunition service to a halt. The wire cable also stretched, and once the cables were no longer of equal length, the platforms were not balanced. When that happened, the floor of the hoist would not match the level of the loading platform and the heavy ammunition truck could not be withdrawn. Locally, there were many examples of the problems. At Fort Flagler, one of the wire ropes frayed and became so separated by blast effects during target practice that it had to be hammered together before it was safe to continue. In 1902 all the lifts of Battery Rawlins were unserviceable, and only one was working in Battery Revere. It was the same elsewhere. At Fort Greble, protecting Narragansett Bay, the lifts in one battery were considered so dangerous that the battery commander had forbidden their use. The officer in charge of one of the groups of forts protecting New York condemned the lifts as "worthless…constantly getting out of repair, and seldom [working] properly." Even G. L. Gillespie, the chief of engineers in 1903, contemptuously referred to the hoists as "almost useless."[42]

In addition to the shortcomings of the hoists themselves, they also seemed to suffer at the hands of troops. If the hoist controls were not operated properly, the cable could spool off the drum, causing the hoist platform to lurch or drop in the shaft, or the winch frames could be fractured or pinion gears sheared of their teeth. So many parts of a single type were broken at Fort Flagler that the civilian engineer employee there recommended that a replacement stock of at least a dozen be kept on hand. Moreover, the flaws in the design of the hoist were exploited unknowingly by the artillery troops. The operation of mechanical devices was an endeavor that invited error, and coast artillery service pointed up the difficulties that even willing recruits had in dealing with appliances that were foreign to them. As one company commander explained in 1903,

> perfection in the use of the lift as designed by its inventor is not claimed for the average enlisted man of this company. The numerous times that these lifts have fallen without the intention of the operators is without doubt due in part to the lift not having been used as contemplated. This was not due to lack of instruction of the men; for in no case have the operators failed to describe minutely the clamps they had sprung that would have prevented the accidents, etc; these safety arrangements provided by the inventor, in general, if used will prevent accidents. But they are not automatic and responsibility rests too heavily upon the enlisted man for him to take the precautions described.[43]

Others supported the idea that the balanced platform hoist was difficult to use. An assistant engineer in the fortifications of San Diego said that he had "seen pawls bent, teeth stripped, keys mangled, cables pulled out and numerous runaways, not through any insufficient strength of parts for the intended use, not through any evil intentions on the part of the boys, but because they did not know or could not think at the time just what to do."[44]

Given such questionable reliability, it is easy to understand why Harry Taylor was willing to go to such great lengths at the Fort Worden main battery to avoid the multiple drawbacks of the balanced platform hoist. He also refused to install them in the eight-inch disappearing gun battery at Fort Ward. There he used simple block and tackle in the vertical hoist shafts. He estimated that a complete charge could be taken from the magazines to the upper level every two minutes. Since there were two shafts, one complete charge could be delivered every minute, adequate to maintain the rate of fire expected from the gun and crew.[45] Taylor's own dissatisfaction with the current devices for moving ammunition, enhanced by his continued involvement in the fortification program after he left Puget Sound, led to his development of a greatly improved mechanical ammunition hoist.

Taylor had not been alone in his wish to replace the balanced platform hoist with something better. He had been preceded by another Corps of Engineer officer who had created a reasonably effective apparatus—the chain hoist—which offered many advantages over the older system. It worked best with light projectiles. Taylor and his colleague, Colonel R. R. Raymond, enhanced the design, making it simpler, stronger, and more flexible. The Taylor-Raymond hoist became the standard mechanical ammunition hoist and was used for all calibers of large guns.

It consisted of a pair of chains moving over large sprocket wheels fixed to two axles. One axle held the lower end of the chains at the magazine level and the other carried the upper end of the chains in the ammunition lobby. Carriers fitted to the chains moved with them as the chains revolved in their circuit over the sprocket wheels, traveling from the magazine to the upper level and down again in a continuous motion.

Each carrier passed through a special platform, picking up a projectile as it moved upward until it reached the ammunition lobby where the projectile rolled off the carrier and onto an inclined delivery table. An escapement at the end of the table held the projectile until it could be deposited on an ammunition truck. There was nothing to get out of adjustment, it was mechanically simple, there were no confusing safety mechanisms and it was easily adapted to all sizes of emplacements. So superior was the Taylor-Raymond hoist that batteries fitted with balanced platform hoists were modified to accept it.[46]

The revision as it was ultimately carried out called for the removal of the concrete separating the two shafts of the balanced platform hoist, creating a single large hoist shaft. In the Puget Sound forts, only one of the two sets of balanced platform hoists was converted, and the other remained unaltered. A single Taylor-Raymond hoist could do the work of two balanced platform hoists, so complete replacement was not necessary. By 1904 work was underway on the ten-inch gun batteries at Fort Casey and Fort Flagler, and the modification was completed in 1905.[47] The same type of hoist was installed in Battery Wilhelm and in Battery Nash in 1906. Taylor-Raymond hoists were also placed in the main battery of Fort Worden during its reconstruction.[48]

The Artillery happily accepted the new equipment. At Fort Ward particularly they had missed the more conventional hoists. The commander of the Artillery District of Puget Sound agreed with them and expressed the opinion that the differential blocks were not a "proper means" of ammunition service by any stretch of the imagination.[49] The men at Fort Worden were so pleased with the new hoists that they began operating them even before the equipment had been transferred to the Artillery, running the hoists by hand since the electric motors had not yet been connected.[50]

There were other modifications as well. In addition to the change in ammunition hoists and the extensions to the loading platforms (at Battery Nash in 1907 and Fort Casey the following year), recesses were cut into the walls of some emplacements for telephones and other communication instruments, iron walkways were added between the first four emplacements of the ten-inch battery at Fort Casey, concrete roofs were fitted over the newly installed Taylor-Raymond hoists, and parts of the parapet at Batteries Revere and Rawlins were removed to permit greater traverse for the gun carriages.[51]

"A battery is like a ship, never finished," offered S. D. Mason, long-time employee of the Corps in the Puget Sound defenses.[52] Refitting emplacements was very much like refitting a naval vessel since both practices extended the useful life of an expensive investment. However, the parallels were limited. Ships occupied so tiny a percentage of the ocean that there was never any worry that there would not be room for them. Fortifications, on the other hand, covered well-defined parcels of real estate selected because of their position on the waterways to be protected.

Properties suitable for coast defense were limited and only a certain number of batteries and supporting structures could be built on those sites. Crowding more construction into already fortified headlands would have restricted the orientation of such new works and would have decreased the overall effectiveness by increasing the density of the armament. The modification of existing batteries then was critical and ensuring that existing batteries were as modern as possible was as significant an enterprise as the construction of wholly new batteries.

The layout of the gun and mortar batteries was a combination of ideas about how a harbor should be defended and those ideas were changing even as construction was under way. The big gun batteries of the Admiralty Inlet forts were positioned as they would have been in the days of muzzle-loading cannon—focused immediately to the front where they could do the greatest damage at short range to any attacking fleet. In that view, volume of fire, essentially many rounds delivered rapidly, was the paramount value. However, artillery officers were changing their understanding of how best to combat vessels as they began to realize that the new weapons now in their possession gave them great increases in range and accuracy, assets that could prove more telling than the ability to fire many projectiles at a target. The greatest contribution to range and accuracy that could be made by fortification design was to provide a firm and steady foundation for each of the cannon of the defense, and that requirement was easily met. Fortification design also had an impact on volume of fire by ensuring that the emplacements were efficient in the way that they aided the rapid movement of ammunition. As we have seen, there were a number of ways to promote that movement and not all of them were successful, the most obvious example being in the original form of the Fort Worden main battery. The engineers learned rapidly. They found better ways to arrange the spaces set aside for ammunition so that less movement was necessary and they improved the mechanisms for getting ammunition to the gun. They also increased the size of the emplacements in new construction so there would be more room for the crews as they served the weapons. All were valuable enhancements but nothing could compensate for the single most critical flaw in the defenses.

What was needed was a collection of guns that pointed west to the Strait of Juan de Fuca and the sea, the direction from which an enemy would come. However, few guns in the defenses could be used in an attack that began somewhere west of Point Wilson. Battery Kingsbury faced almost south and Battery Rawlins faced almost east. Rawlins was particularly limited; neither of its guns could fire to the north effectively and the far right gun was so restricted that it had only a degree or two of azimuth remaining in its traverse after clearing Partridge Point.[53] Battery Ash bent toward the west but it did not cover much water, and the other guns of the Fort Worden main battery faced east.

The poorly planned orientation of the batteries impaired the quality of the defense. No number of batteries, no matter how well designed or constructed they might be, no matter how they might be improved by modifications, could overcome the disadvantage of having been sited to fight a war a generation past. Officers of the Artillery and Corps of Engineers recognized the weak western approach as the central flaw in the defenses. In the years prior to 1915 they would spend much time and energy trying to correct it, years in which the design of naval vessels continued to advance so that warships were becoming less liable to defeat by coastal fortifications.

4
Working Along True Scientific Lines

"The coast artillery [service] has worked out along true scientific lines what is confidently believed to be the most efficient system of coast artillery in existence."
—*General Arthur Murray, 1907*

The Endicott Board of 1885 outlined a unique arsenal. With few exceptions, the weapons it described did not yet exist, nor did the facilities for making them. It called for almost six hundred modern guns of six different calibers and over seven hundred mortars, yet in the United States almost nothing was known of modern ordnance manufacture. The board's proposals relied heavily upon Abbot's unproven mortars, as well as the even more uncertain proposition of the disappearing carriage. The report included turrets and casemates, some of which were in use in Europe, although no domestic firm had the plant or the skill to contrive the same equipment. The naval auxiliaries it embraced were a diverse collection of specialized fantasy, as were the floating batteries. Of all the weaponry proposed, only the submarine mine was anywhere near the stage of development that would permit practical employment.[1]

Despite such improbable beginnings, the armament as installed was ultimately very much like that specified in 1885. The accuracy of the projections lay in the resourceful development of a sophisticated group of weapons and in the ability of American manufacturers to supply the quantities of high grade steel necessary for the construction of the guns. The derivation of the sea coast weapons and their adjuncts is beyond the scope of this work, but since the ordnance mounted in the Puget Sound forts shared origins common to all coast defense armament, some background will help place the fortifications in an improved perspective.

The topic is broader than guns and mortars alone; it includes a cluster of devices that sprang from an era characterized by a creative spirit. They were not without their problems, a few of which became manifest in the Puget Sound defenses. In one instance, a carriage did not fit the emplacement that had been built for it and other carriages proved cantankerous. Although submarine mines were considered essential, they could not be placed in the broad and turbulent waters swept by the guns of the Admiralty Inlet forts, yet they had to go somewhere. Equally important were questions of how to control the large collection of armament and how to best aim the many guns of the many batteries. The answers led to building efforts that paralleled the construction of emplacements for cannon and added important structures to the fortifications.

The new defenses meant new breech-loading cannon, lots of them, in many different sizes. The great numbers necessary for the coast defense program would require large amounts of steel from private suppliers, as well as the expansion of the government's facilities for finishing the cannon. These were more logistical problems than technical ones. More formidable was the need for some device to hold the cannon during firing.

There had to be a carriage—a mounting for the gun—that could absorb greater forces of recoil than had been experienced before, and that could do so without impairing the increased accuracy of the rifled weapons. Additionally, the Endicott Board required that some of these carriages would have to permit the

Plate 4-1. Bringing ten-inch guns ashore at Marrowstone Point, April 7, 1899. Most guns and carriages were barged to the forts after a cross-country rail journey from east coast arsenals and manufacturers. The ordnance was most often put ashore on the beach because the reservation wharves were not strong enough to hold the great weights. At Fort Flagler, a hoisting engine pulled the cannon over temporary grades and trestles to the top of the bluff where they could be maneuvered into the emplacements. *Author's collection*

vertical movement of the cannon to keep them out of view by attacking ships. Other carriages would have to fit in small casemates, and still others within the confined space of turrets. The carriages also had to be able to stand idle for perhaps years and still be ready for instant use. They had to endure the occasional

handling of less than conscientious troops. In battle, the mechanism had to be tough enough to survive showers of splinters and shell fragments. In short, carriages had to be strong and simple, with no "complicated and nicely adjusted machinery."[2]

Plate 4-2. An ordnance mechanic at Fort Flagler looks over a recently installed barbette carriage for a ten-inch gun, a combination unique to Puget Sound. Mounting guns and carriages required expertise, and men familiar with heavy jacks and falls were in demand at all the forts. Sometimes the work could not be done quickly enough, and the job of mounting the cannon (but almost never the carriage) was left to the artillery troops. *Author's collection*

Drawing heavily upon both European precedents and native ingenuity, ordnance designers created an incredible array of devices in fulfillment of the criteria. The dazzling success of one of the requirements of the board—the disappearing carriage—made many of its other weapons unnecessary and gave the coast defenses a character unduplicated elsewhere. On the other hand, the too rapid endorsement of something called the gun-lift carriage was an expensive lesson, and one that had repercussions in Puget Sound.

It took a good deal of courage on the part of the Endicott Board to commit much of the nation's defense to the disappearing carriage. For decades, almost every country with any capacity for producing heavy ordnance had tried to devise a dependable carriage of this sort. The common wish was to achieve some design that would thwart attempts to disable defenses by destroying the crews manning the guns. Increasingly, ships carried machine guns that could sweep the walls of a coastal fort and defeat the gun crews before they could reload their cannon. If a carriage could be devised that could move the cannon below a protective parapet after firing, then the gun could be reloaded in safety, away from shrapnel and small arms fire. A few nations had carriages capable of that, but they were limited to relatively small caliber weapons. The machines were also so encumbered with a profusion of hydraulic and mechanical appliances that they were of doubtful reliability.

In 1893, the Board of Ordnance and Fortification had the chance to examine a new kind of disappearing carriage.[3] It was the creation of two ordnance officers, A. R. Buffington (1838-1922) and William Crozier (1855-1942), and was a dramatic improvement over all other contemporary designs. There were no hydraulic or pneumatic systems or springs to return the cannon to firing position after it had been reloaded. Instead, the force of recoil lifted a heavy counterweight carried on the ends of a pair of levers that extended into a well beneath the carriage. The upper ends of the levers carried the cannon and moved rearwards when the gun was fired. The counterweight absorbed and stored some of the energy of the recoil when the gun levers moved to the rear, raising the counterweight; oil-filled cylinders damped more of the recoil. When the cannon had fully recoiled to the loading position, a pair of pawls locked the counterweight and prevented it from moving down into its well. When it was time to fire the cannon, the gun crew released the pawls, the counterweight moved downward, the levers followed, pulling the gun up and over the parapet into firing position. It was mechanically simple, the average artillery soldier could understand how it worked, and its operation was similar to the barbette carriage adopted a few years before. It became the choice for the service, and with continued improvement, became the only disappearing carriage used in the nation's defenses, mounting cannon of up to sixteen inches in caliber.[4]

The Endicott Board could not foresee the success of Buffington and Crozier, so it looked at several ways of carrying out the disappearing principle. One attractive device was the gun-lift carriage, manufactured in Europe by the French firm of Le Creusot. The gun-lift not only hid the cannon from view during loading, it hid the entire carriage, the complete artillery piece moving up and down through an immense opening in the emplacement.[5] In the end, their performance was poor and the gun-lift idea faded against the consummate success of the disappearing carriage.

Plate 4-3. The cannon itself was the heaviest piece that had to be moved. At Fort Casey, some of the difficulty was overcome in January 1900 when a temporary trestle was built behind the battery. Trucks were placed under the thirty-three-ton barrel and a locomotive pushed it part way up the incline. The mounting crew lashed on a block and tackle, then man-handled the cannon onto the carriage. Mounting guns was a dangerous and tricky business: a man was killed in 1903 at Fort Ward when an eight-inch gun fell while being moved into its carriage. *Author's collection*

However, during the period when it seemed that the gun-lift would be part of the standard armament, the Ordnance Department began the manufacture of three additional examples of the mount. In 1896, two years after the completion of the first gun-lift battery and the decision not to continue with that idea, work was begun to modify the three carriages to make them perform in the same manner as the standard barbette carriage then in service.[6] Once altered, they were scheduled for Puget Sound.

Several individuals dominated the creation of the new seacoast carriages. Among the most prominent was William Crozier. He was a member of the Board on Ordnance and Fortification, and reviewed most of the guns and carriages then being considered. Because of his position, he was careful to excuse himself when the board appraised devices in which he had a personal interest. A major contributor to the design of the disappearing carriage, he had also generated the basic form of the service barbette carriage for heavy guns.[7] These rugged and simple carriages, with their side mounted crane for hoisting projectiles and a narrow loading platform extending to the rear, were intended "to be manufactured cheaply and expeditiously to any extent."[8] As it turned out, most of the big guns would be mounted on the disappearing carriage.

The ordnance developments came together within an amazingly short span of years. The artillerymen who would man the weapons were a little stunned by the suddenness. For decades following the Civil War they had dwelt "in the darkness of the smooth-bore gun and the boiler iron carriage night," using mounts for the cannon that appeared to be "thrown together haphazard, loose in every joint and uncertain in their action."[9] Now, all at once, that was

past. The gun carriages were the most careful flowerings of the artificer's craft, and the guns were now capable of fire so precise it was beyond imagining. "The accuracy of our new type guns is something truly wonderful," exclaimed an artillery officer. "To those familiar with the abnormal results with our present obsolete weapons, it reads like a fairy tale."[10]

Most of the enthusiasm was reserved for the disappearing carriage. To have sought after it for so long and then to have it work so well when it finally arrived was almost too good to be true. One lieutenant, marveling over the smooth, steady, up-and-down motion of the experimental Buffington-Crozier prototype, declared it "a grand artillery see-saw," its action "as gentle and graceful as that of a senorita's fan on a summer day."[11] Like his fellows, he was dazzled by the difference between the new invention

Plate 4-4. A coast defender stands next to a ten-inch gun mounted on a model 1896 disappearing carriage at Fort Casey. The disappearing carriage was a mechanical marvel in its day, and widely admired by those who served with it. According to one observer, it was "as near perfection as could be devised. It is solid, strong, and admirably adapted to the ends in view, and, bearing in mind what it accomplishes and how unerringly and even how gracefully it does it, [it is surprising that it is] of astonishing simplicity, and at the same time striking ingenuity." *Washington State Parks and Recreation Commission*

Plate 4-5. A crew with its ten-inch gun and barbette carriage at Fort Flagler. Unlike the disappearing carriage, the barbette mount did not move, which meant that ammunition had to be hauled up to the narrow loading platform. Noting the difference, one contemporary artilleryman thought that the disappearing carriage crouched down "like a camel to receive its burden," while the barbette carriage required that everything be lifted up to the gun, "like climbing the ladder to the main deck of an elephant." *Author's collection*

and the carriages then in the fortifications. "Turning to horses for comparison," he suggested, "if we let a scrub bronco (disposition included) equipped with a cowboy outfit, represent the old style, [then the disappearing carriage] may quite properly be represented in the same respects by a clean limbed, full eyed, well trained hunter."[12]

Once the plan for a harbor had been approved, work began on the fabrication of the weapons that were to go into it. For each gun and carriage being built, there was an emplacement also

underway. Because of the close relationship between the approved plan and ordnance construction, there was no ability to change the number or kind of weaponry as the building went forward.

Harry Taylor discovered shortly after his arrival in Puget Sound just how limited his options were. He reviewed the plans for the defenses, and noted that they called for barbette guns on Marrowstone Point and Point Wilson. He had some doubts about the choice, and suggested to the Chief of Engineers that they plan for disappearing carriages instead. That would be impossible, came the reply. The barbette carriages scheduled for the two points were already complete, only waiting for Taylor to get on with the construction so they could be mounted.[13]

Taylor encountered a similar circumstance a few years later. He wanted to build all seven ten-inch gun emplacements at Fort Casey in a single act, but he had funds only for four. He proposed to transfer funds from the eight-inch gun battery at Bean Point, a fortification of modest value, and apply them to the more important work at Fort Casey. He found that he could not make the change because of the coupling between battery construction and ordnance manufacture: the carriages for the Bean Point battery were already being assembled and there were no more than four ten-inch gun carriages that could then be assigned to Fort Casey.[14] The availability of ordnance foiled Taylor's plans a third time. In 1900 he had prepared an estimate for completing the final three emplacements at Fort Casey, only to learn that he would receive just enough funds for two positions. Only two disappearing carriages were on hand and there was a model change pending. The third emplacement would have to wait until the new carriage design was ready.[15] No one anticipated what problems that meant.

The armament foreman installing the carriage in the third of the three emplacements that were added to the original four at Fort Casey was greatly puzzled in early 1906 when the pieces did not seem to fit. He had bolted down the heavy traversing ring without difficulty, but when he tried to install the encircling platform plates, he found that the parapet was too close to the carriage. There was just enough room to fit the plates between the carriage and the parapet wall. Not only that, but those plates he was able to install covered the stairways coming from the magazine level. Clearly, the emplacement had been built to fit some other carriage.[16]

The source of the embarrassing problem was cloudy. In April of 1901, the chief of engineers had written to John Millis to tell him that the carriage intended for the final emplacement was a model 1896, identical to all the others in the long main battery.[17] Shortly thereafter, there was a model change with the introduction

of the model 1901, and work on the emplacement stopped for several months until a revised plan arrived in October that showed the proper dimensions and proportions of an emplacement for the 1901 carriage.[18]

Since construction had not progressed very far, all that was necessary was a few alterations in the forms. Building continued and the battery seemed wholly satisfactory until the time came to mount the carriage. There was nothing to do but stop the ordnance work for several months, and wait for the parapet and stairways to be modified to accept the carriage. Hiram Chittenden insisted that the emplacement had been built in accordance with the Ordnance Department plans for the 1901 model.[19] It seems likely that the fault did lay with the ordnance plans. Battery Benson at Fort Worden had been designed for the same model carriage and it too incorporated the identical flaw as at Battery Kingsbury. However, the error there was identified and corrected before the carriage arrived.[20]

Almost without exception, the artillery pieces served successfully throughout their years in the defenses. The altered gun-lift carriages at Fort Flagler were another matter entirely. The original carriages locked in position after the gun was fired; the alteration process removed the locking machinery and permitted the gun to return to normal for reloading. There were other minor modifications to bring the carriage more in accord with the standard service models, but the altered carriages suffered from chronic defects. In 1901 the artillery district commander described the carriages as "a constant source of annoyance," and his dissatisfaction exceeded the abilities of the post ordnance mechanic to resolve.[21] No meaningful work could be done without dismounting the gun, and that would not help stop the flow of oil from a faulty recoil cylinder which had leaked since the gun was first mounted.[22] In 1905, the inspector general called for the removal of the carriages and their replacement by more acceptable mounts.[23] The next year, the loading platform of gun number one was damaged during target practice, rendering the weapon useless; timbers had to be placed under the loading platform to keep it from

Plate 4-6. The twelve-inch gun, like this one on a barbette carriage at Fort Worden, was the most powerful employed in the defenses. This is the same gun and emplacement as shown in Plate 3-4, and displays the changes brought about by the 1906 main battery reconstruction. *Washington State Parks and Recreation Commission*

collapsing and the carriage could not be traversed.[24] Not long after, the entire battery was out of commission, with one carriage "shattered" and the other considered too unsafe to use.[25]

There was a third altered gun-lift carriage in the defenses. In addition to the two at Fort Flagler, there was a single example mounted in Battery Ash at Fort Worden and it experienced none of the difficulties associated with those at Fort Flagler. Nonetheless,

artillery officers were eager to be rid of it. If the carriages at Fort Flagler were so onerous, they thought, perhaps it was just a matter of time before the one at Fort Worden began to deteriorate, leaving a single twelve-inch gun to meet an attack. It was replaced by a new barbette carriage in 1909, and components of the displaced altered gun lift carriage were transferred to Fort Flagler to replace the damaged and broken parts on the carriages there.[26]

Plate 4-7. A twelve-inch gun on an altered gun-lift carriage at Battery Wilhelm, Fort Flagler, fires a service round for the first time on May 17, 1900. The year before, the barrels were resting on timbers when the cribbing accidentally caught fire, heating the steel until it glowed. Although the representatives of the Ordnance Department felt that there was no likelihood of damage, the first firings were made with the gun crew well out of the way. Since the gun-lift carriage was designed for a wholly different kind of role, it was altered to make it more like standard barbette carriages. The most apparent modification was the addition of the loading platform and carriage crane as shown here. *Author's collection*

The repairs helped, but not enough to place the carriages on a par with other barbette mounts. Although they now traversed satisfactorily, they were irksome to set in range. It took the entire strength of one man to turn the elevating handwheel. At that, the gun did not move smoothly, and jerked through its path. The difficulty was somewhat compensated for by the "Belcher Slow-Motion Elevating Device," a home-made mechanism devised by an artillery sergeant at the post.[27] Such ingenious contrivances could not overcome intrinsic faults, and the altered gun-lift carriages remained a compromise at best, an enduring if unappreciated reminder of the nation's early interest in European ordnance.

Guns alone did not defend Puget Sound. There was another weapon, the submarine mine, which was in some ways even more threatening than heavy ordnance. Hidden beneath the water's surface, it was an "unseen and dreaded force" that promised almost certain destruction.[28]

Mines had been used against ships since the Revolutionary War, and beginning in the late 1860s, the Corps of Engineers developed the mine to a high point of perfection. Under the guidance of Henry L. Abbot, the same who had advocated the coast defense mortar, the Corps organized mines into networks called fields.[29] Each mine field was made up of numbers of mines placed at predetermined locations. Electrical cables connected all the mines and ran to a central control station on shore known as a casemate. Troops in the casemate could explode any particular set of mines at will.

The use of electrically controlled mines had many advantages over contact mines, or mines which exploded when struck by a floating object. Since the controlled mines were safe until detonated from shore, there was no chance of a friendly vessel stumbling into the field and being destroyed accidentally. Similarly, an enemy could not use valueless hulks to clear a path because the defense could allow the decoy ships to pass over the mines, saving the submarine charges for a serious attack. Moreover, the controlled mines did not have to be in actual contact with the vessel to do damage. They were fired by judgment, the process

of tracking a vessel through the field until it coincided with the known location of the mines.

The mine itself was most often a steel sphere, thirty-two inches in diameter. It contained a fuse and a charge of explosive; most of the interior was empty, and so gave the mine its buoyancy. An anchor held each mine at a certain depth and prevented it from rising to the surface. A mine had to remain at a level approximate to that of the draft of the vessel it was meant to defeat, and as long as the water was shallow, that requirement was easy to meet. The deeper the water, the longer the line had to be from the mine to its anchor, and the longer the line, the more likely the mine was to move out of its position. If there were swift currents in a deep passage, then a mine defense was not practical since the current would push the mines in unpredictable directions.[30] Admiralty Inlet was both deep and swift, and as a result was not mined. A better location for mine defense was Rich's Passage, the narrow approach to the Port Orchard Naval Station about sixty miles to the south of the Admiralty Inlet forts. It was sixty feet deep as opposed to the depths of more than three hundred feet in Admiralty Inlet, although it did have a stiff current; most of the mines in the project were of an especially buoyant and current-resistant type.[31]

A mine field was a passive defense. An enemy could penetrate an unprotected field by grappling for the control cables or by destroying the mines when they might be exposed at low tide. An adequate mine defense needed gun batteries to guard against vessels which might be lingering beyond the edge of the field, waiting for darkness or fog to cover a dash through the obstruction. The proposed plan for Rich's Passage included a dense collection of ordnance; had they all been erected, more than fifty cannon would have lined a waterway a little more than a half mile wide.[32] Only five batteries took shape, and far less was expected of the Rich's Passage defense than had been contemplated originally. No doubt the reduction was due to the desire to stop the heaviest and most threatening warships at Admiralty Inlet. Presumably those few vessels that might escape would be

Plate 4-8. A crew of the 71st Coast Artillery Company waits to go into action with a six-inch gun at Battery Parker, Fort Casey. This was the smallest caliber mounted on a disappearing carriage, and there was some concern that it would not be able to fire rapidly enough to be useful in an engagement. However, the gun and carriage performed admirably, and the rapidity of fire matched that of barbette mounts with similar cannon. *Washington State Parks and Recreation Commission*

light and swift, and just as subject to destruction by the smaller armament at Rich's Passage.

The mine field was located in two bands. The first extended between Orchard Point and Battery Vinton, passing through a natural obstruction called Orchard Rocks. The second band was about one mile to the north and directly in front of Battery Thornburgh.[33] The mines were placed so that one field would be at a depth appropriate for vessels at high tide and the other for vessels attempting to cross at low tide.[34] In all, more than two hundred mines were scheduled to be planted when the need arose.[35]

A considerable shore establishment was necessary to support a mine field. Besides the casemate, there had to be a storehouse for the mines when they were out of the water, another storage facility for the cable, a magazine for the explosive used in the mines, a light tramway to move the mines about, and a wharf to get the equipment onto the planting boats. Unfortunately for Harry Taylor, mine material began to arrive before he had begun any of the structures.

Early in 1899, Taylor had on hand thirty-one reels of mine cable weighing ninety-one tons and no place to put them. The cable had to remain wet to keep it in proper condition for service, and it was usually stored in a large water-filled tank. Taylor had no such tank. With the cooperation of the navy, he deposited the reels on the beach at the nearby naval station, the winter rains providing an adequate substitute for total immersion.[36]

By spring, he had found a good location for a temporary cable tank. Near Middle Point there was a small stream; appropriately modified, the water course could contain all the cable reels. A crew of laborers widened the stream bed and made it sufficiently deep to provide enough water to cover the reels. At the end of May a tug and scow moved the cable reels from Port Orchard to the new cable tank. From the beached scow, the reels were rolled about 350 feet inland, and then parbuckled down an incline into the tank where they rested on a grid of logs. Once they were in place, a dam was built to impound the water and submerge the drums.[37]

The cables taken care of, Taylor next moved to build a storehouse for the mine cases. He selected a site at Middle Point, not far from the cable tank. Several considerations motivated his choice. He anticipated that other work would be underway on the west side of the waterway since the site was more accessible and would require less clearing.[38] He had determined already to build the mine casemate there, and it was common practice to have all mine-related structures in close proximity to each other. Construction began on the storehouse in April 1900.[39] It was an impressive brick building, rectangular in plan, with tall round-arched windows arcaded along each of its long sides. A prominent bull's eye opening marked the gable ends. It was large enough to accommodate all the mines for Rich's Passage and those for Agate Pass as well.[40] A concrete casemate was completed in 1903, tucked behind a hill of earth and rock.

The mining establishment at Middle Point fell victim to the vicissitudes of the fortification program before it was finished. The mines aside, the only other element of the defenses built on the west shore of the passage was a pair of emplacements for two three-inch guns called Battery Mitchell. The bulk of fortifications and the garrison for the entire post were at Bean Point, on Bainbridge Island. Rich's Passage was an effective barrier to a functioning mine defense since it restricted easy access from the main post. A small launch provided transport from Fort Ward proper, and in this manner a sergeant and two privates traveled back and forth to Middle Point as caretakers. However, the balance of the command was so busy maintaining the armament at Fort Ward that it had no time for drill with the mine equipment. No cables were laid; they all remained on the reels in the cable tank. The engines and instruments in the casemate were not kept in commission. The plant itself could be faulted as well since there was no overhead trolley at the cable tank to lift the reels out of the water nor was there a tramway connecting the different parts of the system. Probably the greatest failing of all was the lack of a wharf. Without it, there was no way of transferring the materiel to

Plate 4-9. A five-inch gun on a balanced pillar mount at Battery Vicars, Fort Worden. The balanced pillar mount was also used for the three-inch guns of Batteries Thornburgh and Vinton at Fort Ward. When needed for action, the gun could be raised on a steel column so that it extended above the parapet. Ammunition had to be handed up to the crew standing on the platform, which made loading slow and awkward. *Dan Kerlee collection*

the vessels that would take the mines and appliances where they were needed for planting.[41]

Some of the limitations could have been offset by more manpower, and the chief of artillery contemplated the assignment of a garrison of eighty-five officers and men to Middle Point. It was an ambitious estimate, not founded in the realities of available troop strength. The only real alternatives were to develop the Middle Point mine plant completely, including a dependable boat service from Fort Ward, or to abandon the site and rebuild the plant at

the main post. A board convened in 1905 to ponder the choice. The brick and concrete buildings at Middle Point were very good and it would not have been possible to duplicate them at Fort Ward except at great cost. If new buildings were to be built at Fort Ward, they would have to be of frame construction. The board prepared two estimates, one for adding the missing elements to the existing site and another higher estimate for new construction at Bean Point. After reviewing the proposals, the board reluctantly recommended that it would be best for the mine service and ultimate economy if a new facility were provided at the main post.[42]

The chief of artillery continued to support the Middle Point site and he urged the construction of a mine wharf to supplement the other structures.[43] As a result of the divided opinion, there was no action taken at either location. The prospects of the mine defense became uncertain. In 1906, a proposal to build a floating dock gained acceptance and although the money was available the next year, nothing was done, probably because of the muddy future of anything at Middle Point.[44]

The matter came to a head in 1908. Battery Mitchell had stood completed for some time, but the guns were ready only in that year. Instead of installing the weapons, the new chief of coast artillery recommended that the guns not be mounted, and transferred to some other battery instead.[45] Viewed alone and without the companion need to support a gun battery, the creation of an embellished post at Middle Point was unnecessary. Work began on the new mine structures at Bean Point.

Finished in 1910, a small group of structures clustered around the approach to the post wharf. Most impressive was the large concrete cable tank, straddled by a traveling crane. On the north side of the tank was the mine storehouse; a tramway equipped with two flat cars linked the storehouse to the wharf, and turntables connected the loading room and cable tank with the main track. The new casemate lay to the northeast of the storehouse. Fashions had changed since the Middle Point casemate had been built. The new casemate was a light frame building, quite roomy, set behind a massive L-shaped retaining wall that was fronted by a gently sloping rise of earth. Not visible from the buildings was the new mine primary station, located high on the bluff near Battery Nash, where observers could track vessels through the mine field.[46]

The effectiveness of the mine defense was compromised by the lack of boats. In 1905 the quartermaster sent a few twenty-foot yawls called "submarine boats" to Fort Ward.[47] A few years later, they were joined by a larger craft known as a distribution box boat, but these were all auxiliaries to the specialized mine planters.[48]

Even when a mine planter was assigned to Puget Sound in 1909, it was shared with the Columbia River defenses and was not always available for planting mines in Rich's Passage.[49] Resourcefulness became the rule. Thus in 1913, when the mine planter *Major Samuel Ringgold* was meeting other demands, the mine command at Fort Ward had to create some method of getting its mines down. Providence provided an old scow, found abandoned on the beach at Fort Worden. It was patched up, made to float, and sent to Fort Ward. It served remarkably well in the stead of the mine planter, although it sank shortly after the exercises were completed. So suitable was the scow that one was specially built for emergency mine service, and put into use in 1915. Unencumbered by the crowded decks typical of most vessels, the crew on the scow was able to plant three mines in twenty-one minutes, which was considered a very good showing.[50]

Admiralty Inlet was never mined successfully, even though it was greatly desired and local commanders were urged to continue experiments in that direction.[51] For a few months in 1898, there was talk of sending a pneumatic dynamite gun to Marrowstone Point. Developed for use in areas where mines could not be laid, the special weapon could throw large charges of dynamite into the water near warships, and was an extension of the idea of the minefield. Because of the estimated cost of the installation, the exotic fifteen-inch caliber weapon never arrived.[52]

Better luck was had placing mines near Deception Pass. The shallow waters of Saratoga Passage inside the Pass and to the front of Goat Island—later named Fort Whitman—were ideal. A single group of nineteen mines were marked for the defense, and a room in Battery Harrison fitted up as a casemate. There were no support buildings on the island; when not in the water, the mines and cables were stored at Fort Worden.[53]

In the 1860s a gunner aiming a smoothbore cannon simply pointed it in the direction of the target and hoped for the best. The new armament of the 1890s required something better,

Plate 4-10. A soldier peers through the open breech of a twelve-inch mortar at Battery Bankhead, Fort Flagler. The prominent spiraling grooves, called rifling, made the projectile spin and that rotation helped make cannon fire more accurate than it had ever been before. As one member of a mortar crew recalled, "it's great to see those big guns spouting flame and smoke and hear the deep, sharp detonation . . . [w]hen they fire in the other pit, one can hear the rushing moan of the bullet like the sound of a strong steady wind thru the branches of a lone spruce tree." *Washington State Parks and Recreation Commission*

since each gun was expected to strike targets miles away, perhaps even out of sight of the battery. Every gun had a telescopic sight attached to its carriage. The gunner peered through the sight, and when the cross hairs centered on the target, he could be assured of a reasonable chance of a good hit at short range. As the distance to the target increased, however, the sight was less accurate, and if the defenses were to approach their greatest potential, a better way of determining the position of the target was a foregone conclusion. For mortar batteries, the need was absolute. Hidden behind hills of earth and far distant from the shoreline, the crews could not see the target from the battery. Without a position finding service, the mortar battery was "as useless as a rubbish heap of sand and iron."[54] Equally important, the officers in charge of the defenses had to have some means of controlling the entire system of fortifications so that any future action would be an organized effort and not a flurry of shells sent indiscriminately at randomly selected targets.

Knowing accurately the distance and direction from the gun to the target depended upon two distinct types of shore-based optical instruments. In what was termed the horizontal base system, an azimuth instrument much like a surveyor's telescope and several operators were required at each end of a long and precisely measured base line, usually several thousand yards. The great length reduced the chance of error in determining the position of the target. Each operator or observer would point his telescope at the target vessel and then read the direction to it in degrees, minutes, and seconds. That information then had to be sent to a central point for processing, and the only possible means of communication was the telephone, which in the 1890s was a recalcitrant and fussy apparatus that could not always be relied upon for the uninterrupted transmission of important data. The angles of observation combined with the known length of the base line produced the basic information about the location of the target. Most crucial to the success of the system was the need for both observers to track the same vessel. If there were only one ship in view, there could be no confusion, but it was likely that any attack would involve a number of vessels, all approaching at the same time. An error by one of the base end stations in the selection of the vessel to be followed would cancel out the correct information gathered by the other station. There seemed to be no way of insuring that both stations would follow the same ship.

There was another method of position finding that had none of these disadvantages. The vertical base system had only one instrument, and instead of a long horizontal base line, there was

a much shorter one: the vertical distance between the observing instrument and the surface of the water. The instrument measured the angle of depression from the horizontal to the target, and was hence known as the depression position finder. That angle, computed with the height of the baseline and the right angle formed by the intersection of the baseline with sea level, provided the distance or range to the target. The direction, or azimuth, came from a scale on the base of the instrument. When it was put into place at each end of a horizontal baseline, it resolved a major shortcoming. The objection to the horizontal base system had been that the widely separated observers might inadvertently follow different targets, making their observations useless. Since the depression position finder could produce readings for distance as well as direction, it was a simple matter for one observer to describe the location of the target to the other observer, confirming that they were both looking at the same vessel. The first examples of the device became available in 1897, two of them arriving in Puget Sound the next year.[55]

The position finders were housed in either protected concrete structures or at the top of frail-looking iron towers. They were collectively known as "stations," in the same manner that groups of cannon were called "batteries." Between 1900 and 1905 the engineers had built four different types of shelters or stations, a variety further confused by changes in how they were to be used and outright cancellation of others that had been started. The contradictory progress was typical. Gun and mortar batteries had received most of the funds for coast defense and little was left for fire control, the term of art that referred generally to the location of targets and the tactical management of the different batteries. To make matters worse, the Ordnance Department, Signal Corps, and Corps of Engineers each had to share in the solution to the difficulty, yet their common responsibilities were funded separately.

There was also no standard issue of plotting boards, ballistic correction devices, and other paraphernalia necessary for computing target information. Some of the materials were made by

Plate 4-11. Fire control structures at Fort Flagler, about 1910. The towers were put up in 1902, and the large building was built a few years later at the time of the Barrancas system improvements. Originally located behind the main gun battery, all have been removed. None of the towers that were once used at Fort Flagler and Fort Worden survive, nor do any of the Barrancas buildings. *Washington State Parks and Recreation Commission*

artillery officers at each post. The 1903 exercises between the Admiralty Inlet forts and the Navy's Pacific Squadron found only the crudest makeshifts for fire control apparatus, and by the next year, the chief of artillery had to admit that almost a quarter of the coast artillery companies were using improvised equipment.[56]

There was also disagreement within the artillery service about what type of system was best, the duties of the battery commanders and fire commanders, the flow of information, and the location of critical elements of the apparatus. One man recalled that at the time, "every officer in the artillery was using a little pet scheme of

his own [for fire control]… Many officers in going to target practice would be so covetous of their special schemes that they would actually hide them from their brother officers as securely as if they were hiding them from the enemy. There were officers all over the artillery working and thinking day and night of some particular system of fire direction."[57]

The subject absorbed the attention of many artillery officers and a great deal of energy went into the planning of a single and universally applicable fire control method. In 1901 a system was put to paper, and it was tested several years later at Pensacola,

Plate 4-12. Two types of fire control station crowd the small hill behind the Fort Casey main battery. In the background are examples of an early design developed for locations where the additional height provided by steel towers was not necessary. The Corps of Engineers improved upon the idea, and in 1903 brought out a standard all-concrete "high site" station, represented by the two structures in the foreground. The small Sewell building to the right rear is the meteorological station. *Washington State Parks and Recreation Commission*

Florida.[58] There, the major contributors to fire control theory came together and fabricated a system which, it was believed, would develop the full potential of the armament. Known as the Barrancas system, after one of the forts in Pensacola Harbor, it had a profound impact on the defenses.

The horizontal baselines of the Barrancas system were different from those that had preceded them in that they were parallel to each other, laid out on the same direction or azimuth, and of equal length. Behind the careful arrangement lay the possibility that during some future battle, one of the base end stations might be destroyed or its communication lines interrupted. Such an event could put a battery out of action, and to avoid that circumstance all

the baselines were interconnected so that one might be switched to compensate for the loss of another. Since the plotting board of each battery was set up for a baseline of a specific length and azimuth, all of the baselines had to have precisely identical features.

One end of every baseline terminated at the battery commander's station. Several battery commanders' stations were clustered around the fire commander's station. The prominence given the fire commander was at the insistence of Garland Whistler. At the time of the tests, Whistler was a major, the rank of a fire commander in most coast defenses. It was his opinion that in time of war, officers of limited experience would command many of the batteries. He thought it essential that the more capable officers (and he was referring to himself here) have the greatest amount of control possible over any engagement, which meant watching very closely the less-skilled junior officers. The fire commander could check on observations and plotting activities, ensure that the battery commander under his charge had identified the proper target, or could even point the observing instrument at the target himself, only if all the stations were close together. The closeness demanded by Whistler was at the heart of most of the structures built in response to the Barrancas system.[59]

Under the plan, the proper place of the battery commander was no longer at the battery itself, but at the battery position finding station because the preparation of fire control information was now considered more demanding than the service of the guns themselves. An underlying belief was that the heavy guns would be manned by militia troops, and it was Whistler's opinion that it would be all they could do to handle the operation of the ordnance. To ask them to have a part in the development of range and azimuth information (as was sometimes done at the gun) would be too much and would probably decrease the competency of the defense. Thus the battery commander would oversee the preparation of the most crucial element, the cultivation of the direction and distance settings for the guns. While there might never be enough regulars to man the guns, it was assumed that there would always be enough well-trained and disciplined men to work the range finders and plotting boards.

The Barrancas system defined for the first time a method of fire control that could be inserted into all the defenses, and determined, also for the first time, the necessary types of equipment for the individual parts of the system.[60] Now each department could make accurate estimates of what the standard system would cost in each harbor. The advent of the system also signaled a major change in the direction of the coast defense program. Since 1890 ordnance and emplacement construction had dominated the appropriations. By 1903 much of that work was slowing, and no appropriations for batteries recommended by the Endicott Board were made after 1904.[61] It was the right time to turn to fire

control as the next step, and in 1905 Congress provided a special sum of one million dollars to begin the orderly and uniform application of the standard system.[62]

The chief of artillery arranged the twenty-seven fortified harbors in the United States in a priority order. Puget Sound was sixth on the list, following installations at the two artillery districts protecting New York, then Boston, Portland (Maine), and San Francisco. It was a challenge to meet the requirements for fire control buildings at all the Puget Sound forts and more so at Fort Flagler. There did not seem to be enough room, or at least enough room in the right place. Some of the officers planning the new system thought that the site occupied by Battery Lee was ideal for a fire control station and they recommended the demolition of the battery. Others held that the loss of a battery with its mounted guns was a step backward. To place the station on the site of Battery Lee, said one, would at the least eliminate a battery from the defenses; it would also place the building in a location subject to attack in foggy weather; and it would suffer from the fire of Battery Rawlins directly over it. On the other hand, to place the station at the other possible location behind the main battery would mean the construction of a building at least twenty feet high in a position exposed to hostile gun fire. The ultimate answer was that the strength of the gun defense was more important, and Battery Lee remained intact.[63]

Because the peculiarities of each fortification made the installation of the Barrancas system more difficult than had been supposed originally, local artillery officers exercised a great deal of influence over the final plan. They were intimately familiar with the reservations, the needs of the batteries, and the water area to be defended. The chief of artillery released Puget Sound from the requirement for equal and parallel base lines, and suggested to the artillery district commander that there be greater freedom in the selection of fire control stations than there had been at other locations.[64] The result was a collection of compromises.

The artillery and engineer officers considering the question at Fort Casey wanted to remove the four existing stations from the

Plate 4-13. A characteristic building of the Barrancas system, this combined primary station at Fort Casey housed the observing instruments for the Fourth and Fifth Fire Commands (collectively, Batteries Seymour, Schenk, Worth, Moore, Parker, Turman, and Van Horne) and the base-end station for Battery Schenk. The ground floor of the building sits in an open excavation, not visible in the photo, with the result that only the tiered observation levels are exposed above the surface of the ground. The photo dates from 1922. *Steven Kobylk collection*

hill behind the main battery because the structures were exposed to fire from several different directions. The problem was that there did not seem to be any more suitable place, and for the time being at least, they would have to be accepted where they were. The officers recommended that the position finding station for the mortar battery, immediately adjacent to the lighthouse, be moved into the emplacements of Battery Turman, and that the battery's five-inch guns be mounted in new emplacements to be constructed some distance away. At Fort Flagler, the examining officers thought the towers to the rear of the main battery too conspicuous to endure and that they should not be retained in any plan. They suggested that new stations be built on the high ground in the vicinity of the existing mortar station near Battery Calwell. Fort Worden presented another set of problems. There

was only one place on the reservation that was suitable for fire control structures: the high ground between the mortar battery and the main gun battery. Here the officers proposed to collect almost all the buildings for the fire commanders and the battery commanders. They tentatively sited two other structures on a high elevation known as Morgan's Hill, already established as one of Port Townsend's residential neighborhoods.[65]

As a result of the scarcity of labor and transportation difficulties at San Francisco following the 1906 earthquake, the installation of the fire control system in the defenses there was deferred in favor of Puget Sound.[66] The sudden availability of equipment and supplies made the completion of the plan for Puget Sound an absolute necessity. To speed the construction, some of the problems of topography that had been so confounding over the years were put aside quickly by acquiring new land for secondary stations at the three Admiralty Inlet forts, and as much use was made of the existing structures as possible. The chief of artillery approved a plan that used almost all the existing stations at Fort Casey and Fort Flagler, and concentrated new construction at Fort Worden. To implement the building program, the chief of engineers ordered Hiram Chittenden east to examine the techniques of fire control station construction at Portland, Boston, New York, and Washington.[67]

The designs that Chittenden produced were typical of the Barrancas system as implemented. In keeping with the wish to place battery commanders in close proximity to their respective fire commanders, the stations accommodated the fire commander and his battery commanders within a single building. The structures had a distinct step-shaped profile, each step representing the observation room level of a battery commander or the fire commander. Each of the Admiralty Inlet forts had at least one of the large combined stations, and Fort Worden had three. To make way for the new Barrancas style buildings at Fort Worden, all the original fire control structures were removed, including the concrete mortar station to the rear of Battery Benson.

Since most of the large caliber gun and mortar batteries were equipped with horizontal base lines, each primary (or battery commander) station had to have a secondary station at the other end of the base line. Two or three secondary stations were often combined in a single building. They were built at the same level and were large enough only for the observer and his telescope, and therefore never reached the impressive scale of the combined primary stations.

The test buildings of the Barrancas system were wood frame structures, since they were to be temporary. Whistler was very impressed with the buildings, believing them to be "only temporary in name," and went on to say that "it would be hard to conceive of a more satisfactory installation for the amount of money expended."[68] Perhaps it was his hearty endorsement that caused the first stations built after the test to be of wood frame. There was

also a degree of concern that a heavily protected concrete station would bring with it the same moisture problems experienced in the batteries, untenable for the delicate optical and electrical devices.[69] Whatever the reason, the permanent fire control system began to take shape as a series of frail and often exposed buildings. Earthen berms and vegetative cover gave a little passive protection, but destruction by fire was an unnerving and greater threat than naval projectiles.

About 1905 John Stephen Sewell (1869-1940) developed a design for the fire control buildings in the harbor of Portland, Maine. It had greater resistance to fire than ordinary frame construction and retained the same degree of dryness and comfort as a wooden building. He used wood frame construction and covered the walls with expanded metal, over which he placed a layer of cement plaster. The technique did not provide a completely fire-proof building—there was still plenty of exposed wood—but the intent was to reduce the fire hazard so that there was little possibility of a general conflagration in a group of Sewell buildings should one of them catch fire. The design was easily adapted to other needs in coastal fortifications, and became common for secondary stations, power plants, guard houses, dormitories, and other small structures.[70]

The installation of the Barrancas system in Puget Sound began in 1907, with most of the required fifty-two stations completed and transferred to the artillery the next year.[71] It was complex and expensive, its cost being among the highest of any of the defended harbors.[72] It was also a distillation of the standard system, with major concessions made in the interchangeability of baselines.

There was some attempt to carry out the creation of identical baselines, and the net result was an odd mixture. For example, there were three separate sets of parallel base lines at Fort Worden, as well as other single baselines that were not parallel to any of the others.[73] At Fort Casey, there were seven horizontal baselines, only two of which were parallel, and two vertical baselines.[74] The same sort of combination prevailed at Fort Flagler

Plate 4-14. The interior of a base-end station. The large instrument is a depression position finder, the key piece of equipment in the vertical base system of position finding. To the left of it are two of the azimuth instruments most often used in the horizontal base system. The inclusion of both types of instruments in the same structure is unusual and it is most probably the result of a training exercise. *Washington State Parks and Recreation Commission*

Beyond the structural disadvantages brought about by the Barrancas system were the adversities worked upon the battery commander. He was so subordinated to the fire commander that he had no real responsibilities. Since his post was at the primary position finding station, he was physically separated from the battery and sometimes located where he could not see it. His value in the primary station was minimal. The fire commander assigned targets, and so complete was his authority that he could even enter the battery commander's station and give what orders he thought were necessary since, under the system, all the personnel of a fire command, even the observers and plotters assigned to specific batteries, were under the control of the fire commander. Once the range and plotting detail were well trained, the battery commander could only stand by, watching the transmission of firing data to the guns. There was an attempt to gloss over the conspicuous omission of the

without any suggestion of parallelism or equality of length.[75] Fort Ward was equipped with a vertical baseline for Battery Nash, although the plan was for a horizontal baseline extending across Rich's Passage to Orchard Point.[76] Interchangeable baselines could work only when all the batteries in a defense had the same field of fire, a condition which almost never prevailed in the Puget Sound fortifications.

The influence of the Barrancas system soon waned and was displaced by simpler means of coordination. Architecturally, however, the buildings erected in 1908 in conformance with its strict principles served for the entire life of the defenses.

battery commander by offering that he could "take such station as the local conditions may indicate to be most advantageous," but it was a hollow response.[77] An officer witnessing an early practice at Fort Flagler commented on the lack of any real position for the battery commanders. During the firing, he noted, they wandered about aimlessly, and appeared to be more or less devoid of any responsibilities.[78]

From the beginning, almost all officers had opposed the separation of the commanders from their batteries. Even the chief of artillery in his comments on the Barrancas system thought the required presence of the battery officer with the fire commander

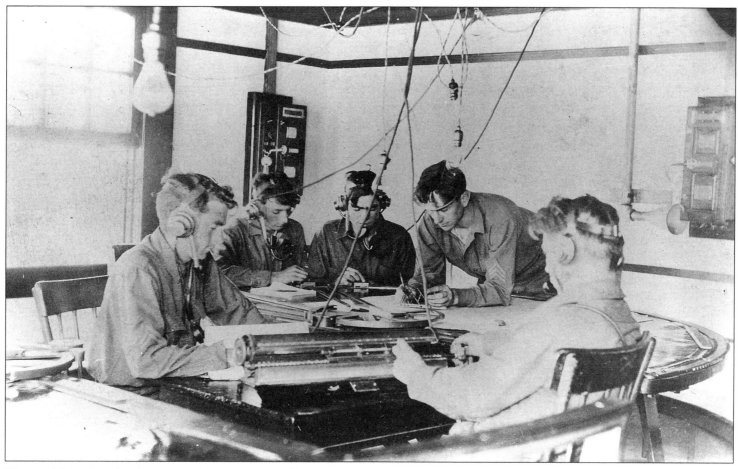

Plate 4-15. At work in the plotting room of a mortar battery. The arm setters (seated, background) are in contact through their telephone headsets with the primary and secondary base end stations, and have set the arms of the plotting board at the azimuths given them. The plotter (leaning over table) has marked a series of observations on the board and is now using an instrument called a predictor to establish the range and azimuth to be used to aim the mortars. He will give his results to the deflection board operator (seated, foreground) who will use his "sausage grinder"—a mechanical computer—to apply ballistic corrections to the data. The deflection board operator will then transmit the elevation and corrected azimuth to the mortar booth at each pit in the battery. The man seated at left is probably a recorder, writing down the results of the plotting for examination later. *Washington State Parks and Recreation Commission*

was a poor idea.[79] The battery commanders themselves clamored for a change that would increase their duties and allow them to oversee their batteries as well as the water area in their charge.[80] By 1912, the grip of the Barrancas system had eased, and they were allowed to return to the guns.[81]

To provide a place for them, a few years later the engineers constructed simple concrete box shelters or stations at the main gun batteries. A single station was placed in the traverse between the guns of Batteries Rawlins, Wilhelm, and Revere at Fort Flagler and, at Fort Worden, the stations were built atop the concrete storerooms behind the main battery; an iron footbridge extended from the storeroom roof to the battery itself. No permanent station for Battery Benson was put up at this time, and instead a wooden shelter was built upon the steep slope in the rear of the battery. The temporary structure was removed in 1933, and replaced with a concrete building.[82] The Fort Casey stations were more elaborate. To raise the battery commander positions for Batteries Worth and Moore above the crest of the two story emplacements, the stations were built as towers. A flight of stairs led up from ground level and a bridge ran from the tower to the loading platform level. Later constructions, such as Battery Kinzie, incorporated the battery commander's station in the structure itself, between the two emplacements.

Just as there were strong feelings about the position of the battery commander, there were equally pronounced opinions about the proper location of the plotting room. Depending upon the kind of aiming method being used, the plotting room could be the center of the position finding activity. The observers at the end of the baseline relayed their azimuth readings to the plotting room where other operators would align the long arms of the plotting board in accord with the sightings. A pencil mark at the intersection of the arms indicated the position of the target at the time the observations were taken at the base end stations. A series of observations taken at intervals—usually about fifteen seconds—produced a series of marks on the plotting board, which indicated the course and travel of the vessel being tracked. The plotting room detail computed the future position of the ship in the terms of direction and range, and sent the information to the emplacement. There the crew traversed the cannon to the proper azimuth and set it at the proper elevation. A pull on the lanyard at the predetermined moment, and the ship and the projectile would arrive at the same place at the same time—or at least that was the anticipation.

Because accurately locating the target was a critical function in coast defense, the plotting room was given maximum protection by incorporating it into the interior arrangement of the first

gun and mortar batteries; Battery Worth is an example of the practice. Like all the other interior rooms in the early batteries, they were often damp and disagreeable. In addition, the noise of the guns firing, the rumbling of the projectile carts on the loading platforms, and the clatter of the chains and trolleys in the magazines all combined to make the work of the plotting room crew almost impossible.

With the advent of the Barrancas system and the centralization of all plotting functions, the plotting rooms were moved into the combined stations. When the requirements of the Barrancas system eased and battery commanders were allowed to go back to the vicinity of their guns, it seemed natural that the respective plotting rooms move with them. Accordingly, new plotting rooms were built at the same time as the new battery commander's stations. Small Sewell buildings in the rear of the heavy gun batteries at Fort Casey and Fort Flagler were completed in 1915, and at Fort Worden, some of the detached concrete storerooms built in conjunction with the main battery were converted to the same purpose.[83]

The same sequence of events applied to mortar batteries, and those at Fort Casey and Fort Worden were supplied with separate plotting rooms to the rear of the pits. At Fort Flagler, artillery officers believed that an unused room in the traverse between the two pits could become an efficient and well-protected plotting room. The plotting equipment was moved out of the battery commander's station in 1914 and relocated to the vacant room, but the reverberation caused by the cannon firing was so severe that the idea had to be abandoned. The next year, all the paraphernalia was taken out and put into a temporary frame building near the battery. The post's commanding officer wanted to build a separate and permanent plotting room like those at the other forts. The engineers were less eager. Added structures were an expedient solution and not preferred over keeping all elements of the battery together. Other artillery units had tried plotting in the rooms of mortar batteries elsewhere and with the same result. However, at Fort Winfield Scott, in the coast defenses of San Francisco, the reverberation was overcome by blocking up some of the ventilator openings leading into the plotting room. The technique was tried at Battery Bankhead in 1916, and most of the objectionable qualities disappeared, eliminating the need for a separate plotting room.[84]

Plotting, observing, and the several command functions developed a considerable amount of information that had to be transferred quickly and reliably to many different users. The flow of data and messages had to occur at any time of the day or night, in any sort of weather, and be resistant to destruction by gun fire. In addition, any means of communication had to be simple to install and easy to maintain. To meet these requirements, many appliances were tested, but the telephone was the most common. Each base end station had at least one telephone so the observers could relay the azimuths to the plotting room. Battery commanders had telephone instruments for contact with the base end stations, the fire commander, the searchlights assigned to the battery (if any), and if he were sufficiently distant, to the battery itself. Fire commanders had a separate instrument for each battery assigned to them, instruments to the several searchlights under their control, and lines to the battle commander and the fire control switchboard. The battle commander garnered an impressive collection of telephones; his station at Fort Worden had thirteen telephone booths on the ground floor. In use at almost any coastal fortification were wall models, desk telephones, special sets for use in plotting rooms, other types for mortar booths, gun emplacement walls, searchlight shelters, and sometimes magazines.[85]

The prevalence of the telephone was unexpected because it had been accepted into the fire control system with a degree of suspicion. The first telephones were not much different from those installed for commercial purposes, and the rugged conditions imposed by coast defense service severely tested the early models. Garland Whistler had little luck with them. He felt that "without any apparent reason therefor they have the habit of tumbling to pieces…"[86] While he was using one telephone, "the head receiver fell off my head, the screw connection and the insulating washers

rolled on the floor, one of the said washers disappeared in an inexplicable manner, and I was obliged during the remainder of the test to hold both the receivers in my hand."[87] Little wonder that some artillery officers feared that fire control entailed "complications that would be embarrassing on the day of battle."[88]

A message sent over the telephone might be lost because of a malfunction in the instrument or noise in the emplacement. To be sure that there was no error in the message, another device was used in conjunction with the telephone. This was the telautograph, an electrical writing machine, and like the telephone it had a receiver and a transmitter. Using the transmitting instrument in the plotting room, the operator could send range and azimuth in a written form that was unimpaired by outside interference. His action of writing the numbers was duplicated by the receiver at the emplacement. Both the telephone and the telautograph were meant to be employed in tandem, with one acting as a check on the other.

Most of the gun batteries in Puget Sound had no convenient place to install either of the instruments. The telautograph was sensitive, and the telephone needed protection from the weather. To house the equipment, small recesses or niches were cut into the walls of emplacements. The work began in 1904, and by 1906 the new communication devices were ready for use.[89] Both the telephone and the telautograph were steadily improved. The later variations of the telautograph were so rugged that they could be mounted directly on the gun carriage. They did require expert care, and there were many instances when they "declined to 'tel aught'."[90] They were abandoned in 1912, in favor of the by then more dependable telephone.[91]

The instruments were meant for a single user, and as long as only one man had to handle the information sent over the line, they were adequate. Matters were more involved at the mortar batteries. Instead of one cannon in an emplacement, there were four, and each had to be set in range and azimuth. The information could not be sent to the men at the guns by telephone since the emplacements were full of the activity of loading and the

Plate 4-16. Mortar or data booths in the rear of the pits at Fort Casey. On the structure nearest the camera in this 1968 view, the rack for the sliding chalk boards is visible, as is the projecting light hood for night use. Ever mindful of the expense of the fortifications, the engineers used window glass salvaged from the old Admiralty Head lighthouse to glaze the openings in the booths. The two larger buildings are the plotting rooms added in 1915, but now gone. *Author's collection*

movement of ammunition trucks, certainly no place for delicate telephone wires. The engineers devised a method of displaying the firing data so it could be seen by everyone in the pit at the same time. In 1905 small concrete structures called mortar or data booths were built in the rear of each pit of the Puget Sound mortar batteries.[92]

They were patterned after a design that had been developed for the defenses of Portland, Maine, and they proved so satisfactory that they became a fixture in most fortifications.[93] The small booths were built of concrete to resist the effects of the cannon

discharge and could be placed quite close to each pit. Along one side was a vertical display of long narrow blackboards arranged in a rack; operators inside the booth could slide the boards into the booths and out again. In operation, the plotting room sent the firing data to the mortar booth, the information was written on the blackboards, and the boards were pushed out where they could be seen in the pit. There was a separate blackboard for azimuth and elevation, and some booths had boards that identified the zone (necessary for the preparation of the powder charge) or listed commands such as "Load" and "Cease Fire."

Mechanical data display, like that used in the mortar booths, had much to recommend it. It did not depend on electricity for its own operation, the information offered was visible immediately, and the data were not as perishable as range and azimuth settings sent over the telephone. Since the information originated in the plotting room, a mechanical system worked best when it could be set in motion by members of the plotting room detail. It was not practical to use the displays in batteries with separate plotting rooms, but in structures like Battery Kinzie, which contained a plotting room between the two emplacements, it was a relatively simple matter. A mechanical display system was part of the battery at the time it was transferred to the artillery.

In Battery Kinzie's plotting room were several large metal discs with numbers painted on the face. The discs served as hand-wheels to put the apparatus into operation as well as indicating the numbers to be sent to the emplacements. Moving the discs also moved a series of cog wheels and shafts that connected the discs to a box mounted on the railing in the rear of the emplacements. Selecting a combination of numbers in the plotting room caused a duplicate set of discs in the receiving box to rotate to the same combination.[94]

The length of the metal shafts made the movable parts very heavy, and it required a substantial effort to start them in motion and considerable strength to stop them in the right place. Another problem was that the numbers were too far back from the face of the box to be seen clearly; unless a person was standing directly to the front, he might misread the figure being transmitted. The chance of error was so great that by the 1920s, the display was no longer used in target practice. As a substitute, telephone lines were installed from the plotting room to each emplacement.[95]

Although speaking tubes connected many areas in gun batteries and in some fire control stations, the telephone remained the most useful device because of its flexibility. Telephones were connected through a central switchboard. Separate lines from observing stations, batteries, plotting rooms, searchlights, from any element that had a part in fighting the defenses, ran to the switchboard. Because it was so completely involved in every aspect of the defense, the destruction of the switchboard would quite possibly mean the loss of an engagement.

The first switchboard shelters in Puget Sound, built about 1908, had been Sewell buildings. Fire control appropriations had lagged, and there was not enough money for better protected structures until 1917, when a deficiency bill, inspired by the war in Europe, provided funds to build new shelters of concrete.[96] The Fort Worden switchboard was rebuilt on its site, and then covered with earth in 1918.[97] In the same year, the lower rooms of Battery Wilhelm at Fort Flagler were enlarged, and the switchboard equipment moved from the building behind the battery into the new facility. A tunnel connecting the underground rooms with the surface plotting room was an afterthought and delayed the completion until 1920.[98] The first switchboard at Fort Casey was adjacent to Battery Trevor. Following the 1917 appropriation, it was relocated to a heavily protected structure that had been built beneath the knoll occupied by the four fire control buildings.[99]

The completion of the switchboards in the years following World War I symbolized the completion of the defenses. They were the last element necessary to make the batteries, fire control stations, and searchlights a unified military organization capable of the task set out decades before. The ordnance mounted in the fortifications and the instruments and systems used to control them were the ingenious products of talented individuals in an age known for its faith in mechanical things. Ingenuity and innovation would remain important in Puget Sound as plans were made to strengthen the defenses in the years before World War I.

Plate 5-1. Emplacement two of Battery Kingsbury, Fort Casey. The battery was vulnerable to attack from the east and the north, and to provide much needed protection, the side walls of both emplacements of the battery were extended to the rear as can be seen to the left in this view. The exposed position of some of the Fort Casey batteries meant that it was a weak point in the defenses, which contributed to proposals to build additional fortifications elsewhere in Puget Sound. *Washington State Parks and Recreation Commission*

5

An Increased and Adequate Defense

"[T]he completion in the extreme Northwest of great railway systems, the rapid development of commercial, agricultural, and manufacturing interests, and the establishment of a navy-yard containing the only dock on the Pacific coast with the capacity for a battleship… press for an increased and adequate defense at an early date."
—*Report of the Taft Board, 1906*

Twenty years after the formation of the Endicott Board, the defenses of the United States had advanced in many ways. Most of the batteries were finished and the major portion of the armament was mounted and ready.[1] While the nation had been busy at its fortifications, however, foreign navies had not been idle, and the ships afloat in 1905 were far more threatening than those of 1885. Although there had been many improvements in military science between those years, there had been even more impressive progress in warships. Pondering the adequacy of the harbor works, an officer remarked in 1905 that there could be "no question as to a change in the most important condition—the target—between the sitting of the 'Endicott Board' and the present time."[2] Just how the defenses could be augmented to meet that condition occupied the Coast Artillery Corps and the Corps of Engineers throughout the balance of years left to permanent defense as part of the military establishment of the United States.

Puget Sound had its own unique collection of problems. Few large caliber guns bore upon the main avenue of approach from the sea. All the Admiralty Inlet fortifications were compromised by Discovery Bay, a neighboring and undefended harbor that could accommodate many invading warships. Frequent fogs obscured the waterway the forts were to cover. The improbability of placing mines in Admiralty Inlet suggested that more forts had

to be built to the south to ensure the safety of the naval station and important cities. The sense of all this was that the early planners had skimped the defenses, and that heavier guns were needed to protect the area, now of considerable commercial and strategic significance. The resourcefulness of local artillery officers resolved some of the difficulties and new construction compensated partially for other imperfections. Other plans to rectify the shortcomings never came to fruition, yet the results of the considerable attention focused on Puget Sound after 1905 demonstrated its prominence among the nation's defended harbors.

The most immediately noticeable failing of the defenses was that the guns were not in a position to attack ships where they were first likely to appear. Fort Worden was the most westward of the three Admiralty Inlet forts, yet few of the guns in the main battery could fire to the west, the direction from which any fleet almost certainly would come. Since the initial and major construction had been concluded before the limitation was appreciated fully, there were limits to the improvements that could be made. A major supplement came in 1904 with the addition of Battery Benson. Shortly thereafter, the power of the main battery was diverted westward by a redistribution of the guns. A ten-inch gun and a twelve-inch gun were interchanged during the rebuilding of the main battery in 1905 so that both of the twelve-inch guns in the battery occupied the most western emplacements.[3] Battery Tolles was also built where it could participate early in an action. That it was of continuing value was marked by the retention of part of its armament during the extensive removals of coast defense weapons for service overseas in World War I, and again in the late 1930s, when weapons were returned to those emplacements vacated many years before. That the western approach required special attention beyond the construction of new batteries was also indicated by interest in Battery Kinzie. After World War I, John L. Hayden (1867-1936), then commanding the Coast Defenses of Puget Sound, suggested that the battery could be improved if it fired more to the west. One of the guns could be swung at an extreme angle if the emplacement were altered by carving out a deep niche into which the cannon could recoil. Because other batteries covered the same area, the modification was never made.[4]

Untouched by the rearranging and the additions was Port Discovery—now called Discovery Bay—six miles to the west of Fort Worden. It was a deep inlet that extended from the Strait of Juan de Fuca, forming a sheltered harbor almost two miles wide and about seven miles long. No guns covered the entrance and it was not visible from any of the fortifications in Admiralty Inlet. When the defenses were being planned, Discovery Bay received scant comment—the intervening elevations were too great to permit naval gunfire from reaching Fort Worden, and troop landings were not considered a threat. By 1904 both possibilities were recognized as very real.[5] Hiram Chittenden summarized the extent of the oversight: "it is doubtful if there is another situation in the world where a great fortification, one which really forms the key to a system of defense, had close in the rear of it, but just outside the range of the batteries, an extensive harbor, perfectly sheltered and large enough to float all the navies of the world, in easy and direct communication both by rail and by good roads and directly on line with the only water supply for the fort and the neighboring city."[6]

Several officers suggested that a mortar battery on the old Fort Townsend reservation could protect Discovery Bay, but there was little prospect for any new construction of that magnitude.[7] More likely was another possibility. Chittenden's appraisal aside, much of the entrance to Discovery Bay was within the extreme range of the mortar battery at Fort Worden. The battery could close the entrance if there were some way of observing the targets. A survey began in 1912 to locate the best site for a remote fire control station on the shore of Discovery Bay, and experimental firings followed the next year.[8] The tests were a complete success, and the special fire control installation was made permanent. Since the mortar battery provided only limited coverage, plans for a more comprehensive defense followed the creation of the Board of Review in 1915.

Attention was devoted not only to the fragile nature of the western approach, but a good deal of concern centered about the core of the defenses at Admiralty Inlet. Prominent in many of the suggestions for the Inlet was an increase in the depth of the fortifications by adding works further up the Sound, an interesting reversal of the more usual seaward migration of coast defenses.

The first to propose the addition of substantial new fortifications was John Millis, who succeeded Harry Taylor as district engineer in 1900. Millis had no previous experience with coast defense design and construction. The greater part of his career prior to coming to Puget Sound had to do with the application of electricity to aids to navigation, the Corps of Engineers at that time being the source of technical expertise for the Lighthouse Board. It was a field in which he excelled. He created the illumination of the Statute of Liberty and traveled to the International Exposition in Paris as a delegate to the International Congresses of Navigation, Electricity and Physics, where he delivered its closing address. He was, in the opinion of one who knew him, "more absorbed in mathematics and other scientific abstractions than in the world of men and women, and by temperament, more of a scientist than a soldier."[9]

After arriving in Seattle, Millis turned his analytical abilities to the question of the fortifications, those completed as well as those yet to be built, churning through the files and drawings that had accumulated in the preceding years. He found much that he did not like. Positions identified for future batteries were the wrong ones, the guns and minefield at Rich's Passage were useful only in a limited way (and not well placed in any event), a new line of fortifications south of Admiralty Inlet needed to be agreed to, and some gun batteries already completed should be moved and rebuilt.[10]

As one product of his review, Millis developed a convincing scenario for the attack and defeat of the forts. A large fleet would approach Admiralty Inlet under the cover of Partridge Point. At night or during inclement weather, several swift and well-armed vessels, accompanied by torpedo boats, would attempt to run by under the guns of Fort Casey, staying in the deep water close to Admiralty Head. The advance fleet would turn into Admiralty Bay, and quickly overwhelm any ships of the defense that might be harbored there. Directing their guns on the fortifications, the invading vessels would harass the batteries at Fort Casey by shelling the unprotected rear of the emplacements, fearing little from the few guns of the defense able to fire into Admiralty Bay. With the forts occupied, the main fleet would force a passage, making good use of rapid fire armament to keep the defenders from their large caliber barbette guns. It would then join the other vessels in Admiralty Bay where the combined forces could begin the destruction of the forts in detail, or go on to other parts of Puget Sound.[11]

Millis believed that to deny the occupation of Admiralty Bay, and to improve the capability of the defenses to fire at targets to the south, heavy guns would be necessary on Lagoon Point, a few miles south of Admiralty Head on Whidbey Island. He did what he could to add the site to the Puget Sound project.

Despite the absence of approval for the idea and repeated instruction to forget about Lagoon Point, Millis persisted. He was convinced that it was the fourth member of a complete system, and he considered the three forts built to that time compromised without it. In commenting on a plan to add other batteries to the three established posts, Millis suggested that they not be located until the works on Lagoon Point were selected. When the assistant to the chief of engineers told Millis that Lagoon Point had no place in the approved project for Puget Sound, Millis wrote directly to the chief of engineers. When the chief of engineers told him that there was no intent even to purchase the property, Millis suggested a reevaluation of all the defenses to demonstrate the importance of Lagoon Point. When that had no effect, he waited until a new chief had been appointed and brought the matter up again. Finally, after nearly two years of effort, Millis relented, insisting that a battery of two twelve-inch guns would have to be added to Fort Casey to compensate for the lack of works on Lagoon Point.[12]

His idea was sound, but premature. Later, the discussions prompted by the recommendations of the Taft Board involved Lagoon Point as a primary feature of the improvements to the Puget Sound defenses. The Taft Board itself did much to emphasize the importance of Puget Sound, and although it brought fewer changes than proposed, its impact was greater in Puget Sound than in any harbor previously fortified under the recommendation of the Endicott Board.

In 1905 Theodore Roosevelt directed his Secretary of War, William H. Taft, to convene a group of officers to examine the report of the Endicott Board. There had been many changes in the intervening decades. The United States had valuable possessions and new naval stations in the Pacific and Caribbean that required protection, the Panama Canal was soon to be under way and would also need the shelter of fortifications. Harbors such as Puget Sound not recognized by the Endicott Board had grown in importance, and warships had become more formidable. In sum, more was expected of coast defenses, and more of them were required to protect broadened national interests.

The report of the Taft Board was an excellent illustration of the growth of coast artillery. When the Endicott Board presented its report in 1886, almost nothing happened. The dire need for the program it outlined received no popular recognition, and the proposals foundered for lack of support. The public was apathetic, some even opposing new fortifications for fear that arming the harbors would be interpreted by potential enemies as inviting war.[13] Congress was little inclined to go where there was no public interest. It did not seem to accept coast defense as an activity of national scope; as one senator said, "no gun has yet been built that can fire a shot from the coast into Kansas."[14]

The Endicott Board plan had to be sold with gloomy conjuring of invasion by unidentifiable foes and even then it was years before Congress appropriated any construction funds. The program probably would have slipped into dormancy had it not been for the spur provided by the Spanish-American War and the determination of professional military men. The Taft Board findings, in contrast, appeared against the background of a firmly established policy of fortification, and its recommendations formed a supplement rather than an entirely new direction. It was well received by Congress and a nation seasoned to the proposition that coast fortifications were a necessary part of national defense. It was more sophisticated than the report of the Endicott Board, and its recommendations were more direct and succinct, virtues made possible by the fundamental success of the program it sought to enhance.

The report appeared in 1906. It included ammunition for the guns and mortars, electrical power plants, fire control systems, and costs of real estate, all items overlooked by the Endicott Board. The Taft Board also gave special mention to mine defenses, and reemphasized the importance of searchlights. More impressive were the proposals for new gun and mortar batteries.[15]

Despite its broad approach, the Taft Board was conservative in that it sought to improve the defenses while staying within the original dollar estimates of the Endicott Board. It asked questions that reflected its bias. The board queried each district engineer about the fortifications and harbors: were the harbors over fortified? Could an adequate defense be had with fewer guns? Did present plans provide too heavy an armament? What was the nature of the excess? Only after these questions were answered was there any interest in new construction.[16] As a result of its investigations, the board dropped two eastern ports from the list of defended harbors.[17] Given the board's incisive nature, its generous recommendations for Puget Sound stood out distinctly.

The board's scheme for the Sound called for the construction of new batteries of fourteen-inch, twelve-inch, ten-inch, and smaller caliber cannon as well as other improvements for the sizeable amount of $5.5 million, a figure surpassed only by the estimate for the completely new fortifications at the entrance to Chesapeake Bay.[18] The board investigated several different methods of intensifying the armament in Puget Sound. The weight of the final improvements rested upon the proposed addition of seven fourteen-inch guns on disappearing carriages. Three were to be emplaced at the northern section of Fort Casey, and two batteries of two guns each were to be located at Double Bluff on Whidbey Island and Foulweather Bluff, on the opposite mainland.[19]

The fourteen-inch guns were a new type, not used previously in the program. For many years, the twelve-inch gun had been the basic large caliber weapon of the fortifications. It was accurate and powerful, and dominated most naval artillery weapons that might be brought against it. However, the energy that the gun developed each time it fired produced high temperatures and pressures that eroded the interior walls, pitting and deteriorating the carefully tooled surfaces which contributed so much to the gun's desirable performance. Ordnance theorists calculated the life of the gun to be sixty rounds, or about one hour and a half in an engagement.[20] The crews fired service rounds at practice, so the actual life of most of the weapons was somewhat less. The guns could be removed from their mounts and relined. With normal use, however, the process would have to be repeated every four or five years, which meant that part of the nation's armament would always be absent from where it was supposed to be, a condition of permanent disarrangement.[21]

The fourteen-inch gun offered an alternative. It was not as powerful, but it made up for the lack of initial muzzle velocity by firing a larger projectile. The larger mass and bursting charge of the fourteen-inch projectile compensated for the greater piecing thrust of the twelve-inch shell. Since the gun developed lower pressures than the twelve-inch model, it had greater endurance, about five and one half hours, or not less than ten years of regular target practice.[22]

The Taft Board endorsed fourteen-inch guns for the protection of wide channels such as Admiralty Inlet; typically most batteries for heavy caliber guns built after the report mounted fourteen-inch guns as a matter of course. They were so desirable that there was some consideration for using them as replacements for twelve-inch guns. In fact, the first designs depicted a compact cannon and carriage which could fit into existing emplacements. Advances in naval ordnance required a more powerful response than could be had with the original fourteen-inch service gun and the next version was longer, making it impossible to use in emplacements other than those built specifically for it.[23]

The board chose Fort Casey as the site for three of the fourteen-inch guns because the position gave excellent coverage to the west. It selected Double Bluff and Foulweather Bluff for an entirely different reason: fog. Instead of covering large areas, fogs in Puget Sound tended to be thick and spotty. Banks clung to headlands or remained in sheltered bays when other areas of the waterway might be completely clear. Quite often, the Admiralty Inlet forts were covered in fog while the area immediately to the south was open and navigable. It was possible for a naval commander to risk passing the forts when they were obscured, and emerge from the fog untouched and unseen by the defense. To prevent that possibility, the board suggested a second line of defense some twelve miles below the original forts. If ships were able to slip through the entrance to Admiralty Inlet, a formidable armament would still be ready to halt their advance.[24]

Congressional appropriations fell short of what was necessary to produce the new batteries, and much of the gun defense proposed for Puget Sound never came about. What did result was the construction of Battery Kinzie at Fort Worden and Battery Harrison at Fort Whitman. For many years, these two batteries were the only fortifications in the United States that could be attributed to the Taft Board.[25]

The genesis of Battery Kinzie did not lie as much with the Taft Board as it did with the artillery officers and their dissatisfaction with the strength of the defense. There had been several attempts at improvement by encouraging an increase in the caliber of cannon in extant batteries. When Battery Benson was under construction, Colonel G. S. Grimes, the artillery district commander, suggested that the defenses could be bolstered considerably if the twelve-inch guns in the main battery were dismounted and placed on disappearing carriages in Battery Benson. The two most powerful guns of the post would then have a much better field of fire.[26]

There was no money to do what Grimes wanted, although another of his ideas had received approval: the consolidation of the twelve-inch guns in the two western emplacements of the main battery. Eager to be rid of the barbette carriages he did not favor, Grimes sought to amplify the modification by suggesting that the emplacements be converted into a battery for disappearing carriages. Several others concurred, but the engineers were less enthusiastic. Changing over old emplacements to accept new ordnance was seldom economical. One estimate for the work was in excess of $185,000, a sum almost equal to the cost of a new battery built from the ground up. The barbette carriages and emplacements remained. Despite his limited success, Grimes had implanted firmly the need for a battery of twelve-inch guns on disappearing carriages, and all the papers covering the subject were gathered up and given to the Taft Board for further study.[27]

The board readily acknowledged the need for the new battery, and included a pair of twelve-inch guns in its recommendations. It did not comment on the difficult problem of location. Fort Worden as a coastal fortification was crowded. Almost every possible building site on the bluff had been used in early construction.

Plate 5-2. Although the Taft Board recommended the construction of new fortifications in many of the nation's harbors, for years Battery Kinzie (shown here at Fort Worden) and Battery Harrison (Fort Whitman) alone represented the products of those suggestions. Battery Kinzie was the last battery built for twelve-inch disappearing guns in the United States, and contained features that represented the culmination of two-story emplacement design for disappearing mounts. *Author's collection*

There remained two potential areas, to the right and to the left of Battery Tolles, yet each was limited in some way. Battery Benson occupied the preferred location. Under the final plan adopted, and actually funded, it was to be demolished, its guns relocated to Deception Pass, with a new twelve-inch gun battery erected in its stead.[28]

The inclusion of the scheme in the Fortification Appropriation Act of 1908 caused considerable stir within both the District Engineer's office and the quarters of the Artillery. Battery Benson had been completed only in 1906, and some thought it one of the best-built batteries on the Pacific Coast. Moreover, it was a very expensive proposition. Estimates to emplace at Deception Pass the ten-inch guns removed from Battery Benson ran to $300,000. It would cost an additional $300,000 to tear down Benson and replace it with a battery for twelve-inch guns.[29]

The proposal was indefensible for reasons of economy alone, and by this time there was also growing sentiment for a large

caliber battery on a low site. During foggy weather, the masts of ships often could be seen from the bluffs at Fort Worden, although their hulls were hidden in the mist. Sometimes the reverse occurred and the fog began above the surface of the water, obscuring the masts, yet leaving the hulls perfectly visible in a

Plate 5-3. The Taft Board brought about the construction of Fort Whitman and its single fortification, Battery Harrison. Pictured is one of that battery's model 1905 disappearing carriage for six-inch guns. It was an improvement over the earlier models used elsewhere in the Puget Sound fortifications, and other than at Battery Harrison, it was mounted exclusively in the defenses of naval bases in the Pacific islands. *Author's collection*

clear belt just below the fog. While the condition was infrequent, it seemed wise to provide for it. The need for a high-power gun battery at a low elevation, when viewed against the breath-taking cost of constructing the same battery on the site of Battery Benson, determined the position of Battery Kinzie on the sand spit at Point Wilson.[30]

As with Battery Kinzie, there was some question about the location of the six-inch emplacements that were to become Battery Harrison at Fort Whitman. The purpose was to protect Deception Pass, a narrow declivity between the mainland and the north end of Whidbey Island. Although the passage was rock-strewn and almost impassable because of tidal currents, artillery officers felt that an energetic enemy would send his warships through without hesitation.[31] Once on the opposite side, there was open and unopposed sailing all the way south.

The Corps of Engineers considered two sites for the defenses. One was on the rocky heights of Whidbey Island overlooking the approach to the Pass itself. The other was Goat Island, a small land mass several miles below the Pass in Saratoga Passage. The chief of engineers selected Goat Island; however, Major C. W. Kutz, then Seattle district engineer, felt that the other location was superior and suggested the plans be changed. Deception Pass was a natural barrier, Kutz said, and any attempt to move through it would surely expose the ships to cannon fire at very close range. He also noted that the state intended to build a steel arch bridge at the site. In time of war, the bridge could be demolished and the wreckage dropped into the water below, restricting the channel further. His proposals made little headway. The chief of engineers replied that the six-inch battery would have to be placed at the rear of the accompanying mine field at whatever location was chosen. Built upon the slopes along the waterway, the battery could be exposed to the fire of an entire battleship fleet, and would be beyond the range of supporting fire from the Admiralty Inlet forts. The field of fire of the battery would be restricted, and the battery itself liable to capture by a landing party. Built on Goat Island, the battery would have an extensive range to the front, and

the water was ideal for mines. The chief of artillery concurred, and the battery was put up on Goat Island, "admirably founded on rock and hardpan."[32]

Another tangible product of the Taft Board was the construction of an elaborate system of searchlights capable of illuminating miles of water. The Endicott Board had included searchlights in its recommendations of 1886. The members of that board apparently considered the lights only for minefields as protection against small expeditions bent on neutralizing the submarine defense under the cover of darkness. The Taft Board went further. It identified the lights as an important aid that would make the defenses almost as effective at night as they were during the day.

The Corps of Engineers had experimented with searchlights since 1872. There were no suitable domestic sources for the lights, and searchlight projectors of any quality came from European manufacturers. By the end of the 1890s, American firms had begun to produce acceptable models, with the exception of the highly polished mirror, which for many years remained the exclusive product of foreign firms.[33] The operation of the lights was simple. A powerful electric current was sent through a pair of carbon electrodes, striking a brilliant arc that was reflected seaward by a parabolic mirror. The diameter of the mirror determined the size of the light, and twenty-four-inch, thirty-inch, and thirty-six-inch projectors were issued to the defenses during the Spanish-American War.[34]

Two of the thirty-inch lights were sent to Puget Sound. They were stored, unused, until 1901, when they were shipped to Fort Ward. These two lights, carried on wagons, were the only night illumination available until the construction of the permanent lights. They were adequate to cover the mine fields in Rich's Passage, but because of the scarcity of projectors, they were used also in exercises at the much wider water area in Admiralty Inlet.[35]

As practice with the defenses continued in the early 1900s, artillery officers began to realize that night was an excellent time for ships to attack. Some felt it was very doubtful if they would ever have the opportunity to engage a fleet in daylight, and they

Plate 5-4. Obscured by heavy woods at Fort Worden, the concrete shelter of Light Four stands with open doors, revealing the sixty-inch projector on its rail car. When needed, the light would be pushed down the track to its operating position at the edge of the bluff. Power for the lights was provided by special generators that were sometimes included in the same shelter as the light. The lights were operated remotely. *National Archives and Records Administration*

drew vivid verbal pictures of desperate night encounters between warships and coast batteries. Understanding the likelihood, the Taft Board proposed that groups of lights be placed in every defended harbor, the number and position of the projectors to be determined by extensive local tests. The lights had to reveal targets at the full range of the guns, which meant that for most locations, especially Puget Sound, the larger sixty-inch searchlight would be the only ones able to fill the special demand.[36]

The War Department adopted the recommendations of the Taft Board and assigned an artillery officer, Capt. W. C. Davis, to command a special "searchlight brigade." Equipped with portable searchlights, Davis and his brigade visited every fortified harbor. They stayed until he was satisfied that they had determined the proper number of lights, and had every one in the right place.[37]

The location of each light required precise positioning. It had to be sited to illuminate the waterway and not any part of the defense. Since it was difficult to see through the beam, the lights also had to be located where they would not interfere with observations from a fire control station or a gun telescope.[38] In Puget Sound, the lights were placed to the side of the most extreme stations; observers could then find their targets within the bracketing beams of the lights.

Even when each site had been selected with meticulous care, the lights were not always easy to use. The actual number that could be employed at any one moment, and therefore the number of ships that could be illuminated, was most often limited to two. Any greater number contributed to confusion in determining which light was on target, handicaps in observing the ship in the beam, and an increase in the accidental illumination of the defenses.[39]

To provide maximum coverage for the wide Admiralty Inlet entrance, Davis recommended sixteen sixty-inch lights, a number exceeded only in the fortifications protecting New York and San Francisco. Two of the lights were to be placed at distant positions, one at Middle Point three miles west of Fort Worden (and not to be confused with the other Middle Point that was part of the defenses at Rich's Passage), and one at Partridge Point, six miles northwest of Fort Casey. These lights, far out from the center of the defenses, would be the first to pick up any approaching vessel. The Middle Point light could also cover the entrance to Discovery Bay. However, the isolated location and lack of protection by fixed guns invited both naval and land attack, and as a result, the lights, numbered One and Seven, were dropped. Lights Two and Three were erected west of Fort Worden, Four and Five at Fort Worden proper, and Six in Chetzemoka Park within the city of Port Townsend. Numbers Eight through Eleven were located at Fort Casey and Twelve at a subsidiary reservation near Fort Casey on the shore of Admiralty Bay. The balance of the lights, numbers Thirteen through Sixteen, were at Fort Flagler.[40]

The engineers constructed a variety of shelters for the projectors. With few exceptions, they were concrete, hidden behind earthen banks or shielded from view with brush and trees to prevent detection during daylight. Many of the lights were mounted on small rail cars; when needed, the projectors were rolled out of their shelters to their operating positions. The typical searchlight installation consisted of the small concrete shelter and another concrete building nearby which housed the generating equipment. On occasion, the generator was remote from the light (for example, the generator for Light Four was at Battery Tolles, about eight hundred feet away), while in other instances a single structure housed both the generator and the light.

Underground chambers at exposed locations on Admiralty Head and Marrowstone Point protected the lights from hostile fire and the blast of nearby guns. At Fort Casey, Lights Ten and Eleven were placed in twin L-shaped pits. The lights were moved out of their protective covering, rotated ninety degrees on a turntable, and pushed some twenty feet further to the face of the bluff. Lights Fourteen and Fifteen, near Battery Lee at Fort Flagler, were set into deep horizontal shafts. Light Fifteen ran out of the shaft and onto a trestle ninety feet long and twenty feet high. The tracks for Light Fourteen crossed over the top of the casemate for Light Fifteen, and included a turntable for maneuvering the light in and out of its chamber. Two of the points selected were low sand spits, and at both locations a tower gave the lights necessary elevation. Light Five at Fort Worden moved vertically through the roof of a former position finding tower and Light Twelve at Fort Casey sat in a small frame shelter atop a timber tower.[41]

The sophisticated installations were restricted to Admiralty Inlet, and the garrisons at Fort Ward and Fort Whitman had to make do with leftovers. The two thirty-inch lights which had been assigned to Puget Sound in 1898 were placed in the outlying defenses, one to each post. A small corrugated iron shelter protected the light at Fort Ward from the weather, and the light at

Plate 5-5. One of the most unusual searchlight installations was for Light Five, near Battery Vicars at Fort Worden. A fire control tower, built originally on the bluff behind the main battery, was dismantled and erected again on the Point Wilson sand spit. The roof was rebuilt with a sliding hatch, and the transformation was made complete by placing a searchlight in the former observation room. The hatch could be moved back and the light winched up until it rose through the roof. *National Archives and Records Administration*

Fort Whitman enjoyed a similar accommodation. It was kept in the warehouse on the wharf. A doorway was cut into the wall, and through it the light was rolled out on a pair of planks, control and power cables trailing over the rough board deck.[42]

Plate 5-6. One of the two searchlights on the bluff at Fort Casey, shown midway between its shelter and its operating position. These lights were kept in casemates recessed into the ground; they were visible only when they were shifted to the face of the bluff. Each projector produced 180,000,000 candlepower. In 1913, the captain of a vessel claimed that the brilliant display of lights from the forts so distracted him that he ran his ship aground at Point Hudson. *Washington State Parks and Recreation Commission*

The searchlight system, turned over to the Artillery in 1911, remained unaltered until the late 1930s. By that time, the wooden piers in the tower of Light Twelve had become so deteriorated

that the light, and later the tower, had to be removed. All of the lights at Fort Flagler were renumbered as a result. A more significant change took place at Fort Flagler with the modernization program brought about by the advent of World War II. Light Fourteen was removed from its exposed operating position on the timber trestle and relocated immediately to the front of Battery Lee. The light rose out of an underground room on a hydraulic lift—and was thus a disappearing searchlight—and was one of the few so mounted in the United States.

Aside from the searchlights and several gun batteries, other marks were left on Puget Sound by the Taft Board, though of a less material nature. The board had brought about the first careful examination of the competency of the defenses, and its report was an interpretation of how they might be improved. Scarcely had the board published its findings when others began to offer their ideas in response, beginning a long period of evaluation that reflected changing views of coast defense.

The Taft Board had not dealt satisfactorily with the basic failing of the Puget Sound fortifications: the western approach from the Strait of Juan de Fuca. Although it had proposed seven fourteen-inch guns, it placed the bulk of them south of Admiralty Head, behind the line of existing fortifications where they would be no use at all in the early moments of a battle. Hiram Chittenden suggested in 1907 that the board's proposals be altered to create a better solution.

He advocated moving the fourteen-inch guns to the three Admiralty Inlet forts, throwing the heavier weapons seaward of the Taft Board locations. One battery at Fort Casey, to be built either at the north end of the reservation or upon the site at Battery Turman, would be for two guns only. He planned another battery at Fort Worden on the sand spit, where it would be in an excellent position to fire toward an advancing enemy. He proposed one more pair at Fort Flagler, making a total of six in the defenses, one less than the number recommended by the Taft Board. He retained the board's reference to the planned relocation of ten-inch guns for the Deception Pass defense, advising the removal of the guns from Fort Casey rather than Fort Worden, and the replacement of those weapons by twelve-inch guns.[43]

Chittenden's plan did not include the batteries proposed for Double Bluff and Foulweather Bluff. The two points were out of range of supporting fire from the Admiralty Inlet forts, leaving a gap between the inner and outer defenses that was not covered by any armament. A fleet able to pass through Admiralty Inlet would then have a safe zone in which it could reorganize to attack the second line of defenses. Also, if ships had slipped by the first line under the cover of fog, they could wait securely until the fog obscured the second line, and move past it as well. Chittenden recommended that the inner line be shifted to Nodule Point, below Fort Flagler on Marrowstone Island; and Bush Point, on Whidbey Island opposite Nodule Point. The locations were close enough to existing guns so that there would be no area unswept by the defense. Some of the ability to adjust for local fogs would be lost at the new sites, but Chittenden believed that the increased tactical value more than outweighed any such disadvantages. He urged twelve-inch guns for Nodule Point and Bush Point, as he did for Lagoon Point. Chittenden was impressed with Millis' earlier appraisal of the importance of Lagoon Point, and planned a special battery there with the express purpose of eliminating any potential shelter for a hostile force in Admiralty Bay.[44]

Early in 1908 Colonel F. V. Abbot, assistant to the chief of engineers, stopped by Chittenden's home (at this time Chittenden was quite ill), and commended him for his ideas, Abbot telling him that they had his approval. As a result of Abbot's influence, the Taft Board project for Puget Sound was modified slightly by 1910 by way of a compromise that sought to resolve many of the shortcomings of the defenses. The fourteen-inch gun battery at Fort Casey remained in the plans to extend the westward reach, and was returned to its original three guns. To close Admiralty Bay, two fourteen-inch guns and a mortar battery were added to Lagoon Point and to provide greater depth, another battery of fourteen-inch guns was scheduled for Nodule Point.[45]

A few years later naval vessels had made additional advances. By 1914 capital ships mounted fifteen-inch guns that exceeded the range of the heaviest weapons in the fortifications, and the numbers that could be massed in a fleet were greater than the guns of the defense. There was some consolation in the superior accuracy of shore-bound artillery, although any sense of wellbeing was brought up short by the ability of modern warships to bombard the defenses from outside the limit of the forts' guns.[46] In response, the secretary of war created in 1915 the Board of Review to revise the projects of the Taft Board.[47] Like the Taft Board and the Endicott Board before it, the Board of Review recommended significant changes in the coast defenses.

The board designated sixteen-inch guns and mortars as the new standard heavy weapons. With ranges in excess of thirty thousand yards, they were the only artillery pieces that seemed likely to match the potential of naval ordnance. The board also proposed a special carriage to permit long range firing with other calibers, and also pointed to the utility of portable guns on railway mounts. It sponsored the equipping of the fortifications with anti-aircraft weapons to meet another new threat, and proposed the abandonment of some older and limited armament. The recommendations presaged a new epoch in American coast defense. In Puget Sound, the board brought few real changes, although it did forecast in some ways the final shape of the defenses.[48]

The board endorsed the importance of works on Lagoon Point, and devised for the first time an adequate protection for Discovery Bay. Lagoon Point was to be armed with a single sixteen-inch gun, a pair of six-inch guns, and a battery of eight mortars relocated from Fort Casey. A pair of sixteen-inch guns in the company of batteries of six-inch and smaller guns, as well as a mortar battery on the Fort Townsend reservation, were to close Discovery Bay. Another sixteen-inch gun was projected for the northern section of Fort Casey on land originally purchased for the construction of the fourteen-inch gun battery recommended by the Taft Board. All the sixteen-inch guns were to be mounted on disappearing carriages able to traverse a complete circle, which increased materially the land and water area covered by the projected fortifications.[49]

The plan was changed slightly in 1916. The batteries at Lagoon Point were dropped in favor of a lighter armament on Foulweather Bluff, to have been made up of weapons transferred from Fort Casey. The armament proposed for Middle Point was unaltered, and a second sixteen-inch gun was added to that already suggested for Fort Casey. Additionally, the caliber of the mortar battery recommended at Old Fort Townsend was changed from twelve- to sixteen-inch.[50]

The design was again changed the year following. The works at Middle Point remained—two sixteen-inch guns, a battery of three six-inch guns, and a battery of two three-inch guns—supplemented by four sixteen-inch mortars. The guns planned for Fort Casey were moved to Partridge Point in company with two six-inch guns and another battery of sixteen-inch mortars.[51] There were further revisions in the proposals throughout the 1920s and '30s, all without result until 1943, and the construction of emplacements for two sixteen-inch and two six-inch guns on barbette carriages (batteries 131 and 249 respectively), not at Middle Point but Striped Peak, west of Port Angeles; and of a six-inch battery (battery 248) at Partridge Point.[52]

The modifications, both real and proposed, reflected a number of departures from longstanding practice. One of the most notable was the reduction in the number of calibers considered necessary to protect any one harbor. The defenses constructed under the proposals of the Endicott Board employed ten different sizes of cannon, each of which required its own ammunition, emplacement, and separate equipment.[53] There seemed to be no good reason for the considerable armament inventory; it made supply difficult and any homogeneity of the system impossible. As early as 1903 there was regret that the guns had not been restricted to only three calibers.[54] The Taft Board included fewer weapons, and the 1915 Board of Review set the pattern for the future by basing its defenses on sixteen-inch and six-inch guns alone, although the proposals for some years carried occasional references to

three-inch guns. The removal of coast artillery weapons from their emplacements during World War I rapidly and unexpectedly accomplished in fact that which had been discussed previously only in theory.

The 1915 Board of Review broke with the past in another sense. Until it made its recommendations, almost every individual who contemplated the needs of Puget Sound had indicated a special necessity for other defenses inside the line formed by the Admiralty Inlet forts. The reasons were one of two: to increase the depth of the defense to compensate for the absence of a mine field, or to adjust for the obscuring effects of local fogs. In both instances, the defenses were perceived as a barrier, and the quality of the barrier was evaluated in terms of its relative impenetrability. The barrier served to separate an enemy fleet from commercial wealth and military resources, the familiar orientation of the Endicott Board that was also accepted to a degree by the Taft Board.

The Board of Review discarded the logic, and selected an entirely new perspective. By throwing the major caliber batteries westward and away from an inner line, it provided a very large expanse of water that would have been protected from enemy bombardment. The extent of the water area was important because a friendly fleet could now deploy for an attack from behind the shield provided by the guns.[55] The fortifications thus acquired a strictly military value which had nothing to do with the vision of fixed fortifications as some sort of national insurance policy.

The idea was so irresistible that it became something of a fashion to extend the protective line as far seaward a possible. The Board of Review established the line between Middle Point and the northern extremity of Fort Casey; later changes pushed that end further outward to Partridge Point. The Chief of Artillery suggested an even greater defensive arc by proposing a series of fortifications on Middle Point, Smith Island (a few miles west of Whidbey Island), and the Deception Pass vicinity.[56] The ultimate fulfillment came in World War II when batteries at Striped Peak and Partridge Point, in conjunction with the Canadian defenses at Victoria, closed off a section of the Strait of Juan de Fuca forty-eight miles west of the original fortifications. Construction of another group of batteries near Cape Flattery, the westernmost mainland extremity bordering the Strait, would have secured the entire water system. Work began upon them, so far removed from the original forts that they were to have been given a new and separate harbor defense designation, but the satisfactory progress of the war brought the building to a halt.[57]

The desire to improve the fortifications, a wish that dated almost from the time a laborer turned the first shovel of earth, was prompted by the early underestimation of the requirements for local defense. Puget Sound had been the orphan of the Endicott Board, and when finally included in the program, the defenses were miserly in comparison with other locations. Although the harbor entrance was wide, there were only a few twelve-inch guns, and all but one of those mounted on the altered gun-lift carriages that were in disrepute before they left the founders' shop. Officers of the Artillery and the Engineers understood the manifest need for a better defense, as did the various boards that tried to strengthen them. That need was dramatically outlined in the years following the Japanese victory in the Russo-Japanese War, when many military pundits believed that a war between the United States and Japan was soon to come. To prepare for the anticipated conflict, the chief of artillery in 1907 sought to increase the ammunition available in all the Pacific Coast fortifications, even to the extent of diverting shells and powder from defenses on the east coast. Leading his list of priorities were batteries at Fort Casey and Fort Worden.[58]

It was the keen sense possessed by both Hiram Chittenden and John Millis that saw the dilemma most clearly. Chittenden felt that the practice of fortification engineering was more exacting in Puget Sound than in any other location. "It is more necessary here than in most other situations," he said, "that the guns, carriages, and other equipment be as good as the best that are being supplied at even such important points as New York and Boston."[59] It was too late for that; early neglect could not be overcome. However, the many plans for improving Puget Sound made it clear that the defenses were among the most demanding. In fact, had the suggestions of the 1915 Board of Review been realized, the collection of batteries in the Sound would have become the nation's largest single complex of coast defense.[60]

Endowments of Knowledge, Skill, and Determination

"Guns and forts in themselves are poor harmless masses, whatever their inherent power and strength, unless they become endowed with the knowledge, skill, and determination of well drilled and instructed men."
—*Lieutenant John Ruckman, 1894*

As the fortifications of Puget Sound took shape, the artillery troops that would make the fortifications of Puget Sound function began to arrive, even as construction continued around them. The close contact between the business of building and the practical application of manpower, which would turn the fortifications into a real deterrent, at times resulted in sharp words and misunderstandings. That was soon past, however, as the new posts developed into small towns that were attractive additions to nearby communities such as Port Townsend. That small city became a magnet for the soldiers in the Admiralty Inlet forts and the fortifications proved an equally compelling curiosity for its citizens. As the defense expanded, the army added more men and training began in earnest. Deep reports of heavy guns soon echoed in Puget Sound. The defenses demanded a supply of artillery soldiers that the regular army could not provide and soon the militia (or National Guard) became a necessary partner, providing gun crews and being called upon to assume responsibility for the landward protection of the fortifications.

Despite what should have been a cooperative alliance, the Artillery and the Corps of Engineers were often in conflict during the years of transition when construction was still active. The running dispute was characterized by a lengthy series of petty incidents which, while not of substance in any one instance, collectively fostered a fractious relationship. Officers and civilian employees of the Corps dutifully recorded the mishandling of equipment and damaging acts on the part of the artillery troops, and just as earnestly, artillery officers noted the shortcomings of the engineers. The quibbling began as soon as the posts were garrisoned and did not cease until construction work stopped.

What exasperated the Corps of Engineers was the Artillery's seeming lack of concern for the batteries that the Corps had labored so hard to construct. Diagrams outlined the care the batteries required, and hung prominently in view, but the troops did not always pay much attention to them. Called to investigate an inoperative drain in Battery Rawlins, the assistant engineer at Fort Flagler found the drain jammed with waste, blue clothes, rags, dirt, and broom splinters, all deposited there, he surmised, by soldiers seeking a handy disposal for their rubbish.[1]

To reduce the possibility of damage when the guns were fired, the Corps advised that battery doors be left open. Time and again, doors were sprung from their hinges during firing because they had been left shut. In one instance, a set of doors to the powder magazine in Battery Powell was blown completely off because they had been shut, and the lavatory stalls in the latrine destroyed through the same cause.[2] At another battery, soldiers forgot to place a stop on a trolley rail and then inadvertently ran the one-ton hoisting block off the end of the rail, splitting the case and springing the gears from their bearings.[3]

It wasn't always an issue of damage or neglect. The engineers had designed the batteries with careful consideration of how they were to be used, and in the plans they had designated specific rooms for specific purposes. It was galling when the artillery troops used the rooms otherwise. In 1903 John Millis noted that in the main battery at Fort Worden, a guard room was being used to store small arms targets, another guard room held the crating for the commanding officer's private furniture (the commanding officer lived nearby in one of the large bungalows built by the Corps of Engineers until his own quarters could be finished), and other carefully labeled spaces were put to such unplanned utility as the storage of small arms ammunition, paint, wooden blocking, and brooms. Only the latrines seemed to be true to their original function.[4]

The criticisms made by the Artillery were more general and often more personal. Harry Taylor wrote to Captain J. D. C. Hoskins at Fort Flagler and asked him to return a sum of money collected from the operator of a boarding house on the post; the money was for the use of water, but the water was from a well built by the Corps of Engineers and the keeper of the boarding house was entitled to use it. Hoskins became very angry and called Taylor's bland text presumptuous, offensive, and impertinent. He added an endorsement to the letter, making clear that he believed that "its sound is not unlike that of some other letters received in the past from Captain Taylor."[5]

A few years later at Fort Worden, a surprised W. T. Preston discovered that Garland Whistler, the artillery district commander, had made formal complaint against him to the district engineer. Preston stood accused of using insulting language to the post adjutant and of taking an uncooperative attitude toward post regulations. The complaint was founded upon an episode in which a water wagon, needed for the construction of a distant searchlight power house, was not admitted to the post because of

an order which restricted the use of teams on the Fort Worden reservation. Preston had not heard of the order before and called the post adjutant. He did not feel that he should call Colonel Whistler since Whistler had issued an order forbidding anyone to call him except in a great emergency. There had been no insulting language, Preston reported, neither on Preston's part nor on that of the adjutant. Preston had explained to the adjutant that he had to have teams come in on occasion or he could not continue the work. In whatever manner Whistler received the retelling of the event, he was provoked: he demanded an apology and if one was not forthcoming, he determined to refer the matter to the War Department. Preston agreed to call on Colonel Whistler, not to apologize he said, since there was nothing to apologize for, but to straighten matters out.[6]

An artillery officer making an inspection of Fort Flagler complained of noxious smells originating from the Corps' reservation. The assistant engineer at Fort Flagler received the complaint and smarted under the attention. He commented to the district engineer that it appeared that the "olfactory organ of the inspecting officer" seemed to have been especially trained to detect odors relating to the engineer department and were "entirely oblivious to the stench arising from the excrement and garbage deposited by others." Who the others were he made pointedly clear, namely the men of the 94th Coast Artillery Company who kept, close to the barracks, a pig pen which produced a reek plainly noticeable at one hundred yards.[7]

The criticism was wounding, particularly for the engineers. The Corps counseled its officers to avoid circumstances that might provoke the Artillery. For example, construction delays were often described in some detail, but by 1902 the practice was discouraged because the explanations were often made the grounds for "captious criticism" of the Corps' work.[8] Hiram Chittenden occupied the position of Seattle district engineer at a time when the tide of animosity toward the engineers had reached its height. He unwittingly inherited the resentments of several years' standing. The difficulties became so pronounced that a friend in the office of the chief of engineers offered Chittenden some advice. The counsel reflected the attitude of the Corps toward the Artillery and incidentally defined the Corps' position in the development of the defenses as a whole:

> We do not want to put ourselves in the position of a superior and unapproachable power, for we can not hold such a position—we would not be supported for a moment, and should not be. Our one object should be to combine heartily with the Signal Corps, the Ordnance Department, and the Artillery to advance the interests of the defense as far as our combined efforts and appropriations permit.
>
> The relationship of District Engineer to the Artillery is extremely difficult to fill. At Fort Casey and the other forts in your district, the not distant past is full of disagreeable endorsements and papers. If by your open and free discussion with the Artillery of all subjects in which you are mutually interested, you can establish a feeling of cordiality and mutual respect, you deserve a maximum on your recitation and a 'transfer to the first section,' in old West Point phrase.
>
> We don't acknowledge the right of the Artillery to dictate but where several courses are open, we usually confer with them and adopt the method that is most accordant with their ideas, or best subserves their convenience, if the cost is not thereby increased.[9]

It was practical wisdom which Chittenden accepted immediately. The next year, he told an assistant that "it is not advisable to endorse too readily the plans of the Artillery for things that it is impossible for us to do, as it simply leads them on to expectations which must result in disappointment."[10]

Some artillery officers advanced the notion that the Corps of Engineers should be eliminated entirely from the planning and construction of seacoast fortifications, those functions to be absorbed by the Artillery, but in the main, the relations between the two branches steadily improved in the decade preceding World War I.[11] They gained a far more effective understanding of their mutual concerns. One consequence in Puget Sound was the

shared discussion inspired by the revisions of the defenses based upon the recommendations of the Taft Board. The collaborative effort would have been difficult to imagine several years before.

Those tense conversations took place between the officers of the two branches of the army most involved. For the ordinary soldier, his introduction to the new Puget Sound fortifications was more gritty and offered few opportunities for reflection.

There was little to cheer the first troops arriving in Puget Sound. On an early September evening, 1899, the officers and men of Battery B, 3rd Artillery, landed on the beach near Marrowstone Point.[12] They moved upward to the top of the bluff near the as-yet weaponless main battery and set up about two dozen small tents organized in neat rows. This was to be their home throughout the winter rains, located in a small clearing on the tip of a forested island untraveled by anyone save the few settlers who tried to farm there. A small detachment soon moved to Admiralty Head and took up equal quarters. There the tents were damp and pounded so in the frequent squalls that it was difficult to sleep.[13] Instead of artillery drill, the men found simple drudgery. They struggled with ropes made slippery with mud to mount the remaining mortars and labored in the rainy darkness to unload the ten-inch guns from the barges beached on the shore of Admiralty Bay.[14] There were few amusements other than drinking, although the lieutenant in charge of the Admiralty Head group found some diversion in hunting rats.[15] A few years later when troops arrived at Fort Worden, they too were housed in tents, but all the soldiers were soon shifted into temporary frame barracks. Little more than plain board shelters, they would serve until the completion of the large and stately permanent buildings that would be home for the coast artillery soldiers.

With the exception of Fort Whitman, which was garrisoned by rotating small detachments from Fort Worden, all of the posts had a complement of well-built one- and two-story structures and certainly the grandest assemblage was at Fort Worden. It was the largest of the Puget Sound installations and its position adjacent to the city of Port Townsend seemed to emphasize its

Plate 6-1. Garrison buildings behind the Fort Casey main battery, about 1903. The large rough lumber building on the left is the temporary barracks, which replaced the tent troop housing on the same site. Other structures in the photo include the homestead cabin used by the engineers as their Admiralty Head office. Almost all of these buildings were removed by 1905, when permanent frame structures were put up some distance away from the gun batteries. A large portion of the site in the photograph is now occupied by a visitors parking lot for Fort Casey State Park. *Washington State Parks and Recreation Commission*

elegance. The community thought well of the improvements, the local paper claiming it to be "perhaps the prettiest post in the country" and that it had "more the appearance of a metropolitan city than a coast defense fortification."[16] For the three forts at Admiralty Inlet, Port Townsend was the focus of commerce, recreation, and transportation to other destinations. As a seaport, there were plenty of opportunities for single men to find entertainment, whether a drink in a bar, a game of pool, or something else in a part of town designated as the "restricted district." Payday was a busy and happy time as the enlisted men looked for ways to spend their money, and to help stretch the limited cash available to them as well as to curb their possibly wayward

morality, the YMCA in 1912 established a service club downtown that offered inexpensive hotel accommodations, pool tables, and a reading room.[17]

There were other diversions; fishing and hunting were popular with officers and enlisted men alike and the larger posts each had a gymnasium. There was an auto club at Fort Worden, a dramatic club, a brass band composed of both soldiers and civilians, and several teams for baseball, basketball, and football. For the officers and their families, there was a succession of balls, dances, card parties, cotillions, teas, and performances, almost all of which involved Port Townsend citizens; attendees from Fort Casey and Fort Flagler traveled to the events by steamer. The connection

Plate 6-2. A group of masons at Fort Casey. In 1900 at Fort Flagler, and in 1904 at Fort Casey and Fort Worden, work began to erect large and spacious permanent buildings to replace the ramshackle temporary quarters built in the early days of troop occupation. Permanent buildings at Fort Ward followed in 1910. After the decline of the defenses, many of the buildings could not be maintained, and in 1937, a large number of the elegant structures were removed from Fort Casey and Fort Flagler. *Washington State Parks and Recreation Commission*

between the posts and the surrounding area was close. Residents of Port Townsend could usually attend movies at Fort Worden, and the commanding officer at Fort Casey hosted a barbecue in the balloon hangar in celebration of Armistice Day, and also opened the reservation to the American Legion for a Fourth of

July celebration and to the United Spanish War Veterans for one of its ceremonies.[18]

The fortifications themselves were a great attraction from the very start. W. T. Preston reported that every Sunday in the summer of 1898 sightseers came out to Point Wilson in swarms, most with cameras in hand. There was little concern at the time since there was almost nothing to see, but as construction continued, the public was excluded from the work site and, ultimately, the entire reservation. From time to time, an individual would apply to visit the defenses, such as the young man and his lady friend who wished to use the batteries as subjects for her sketchbook, or the minister who sought entrance as a prize in a church raffle, but they were almost always turned away. A lodge organization did receive admittance for a few of its members, no doubt because of fraternal ties with one of the assistant engineers. Officially, visits were forbidden until 1906, when the policy was relaxed, yet even as late as 1915 civilian visitors found all the apparatus of the defense a "decided novelty."[19]

An attraction for residents and visitors alike was annual target practice, the time each year when the entire defense system was put through its paces and the big guns were fired. People were cautioned to keep their windows and cisterns open to prevent breakage, and at Port Townsend especially many gathered on the bluffs above town to watch the action. Night drills were deemed especially entertaining, with the play of the searchlights and the course of projectiles marked by tracers. The discharge of the cannon at Worden and Casey was not a particular disturbance, since the guns faced away from the city, but it was a different story when the "big howlers" at Fort Flagler came into play and the shock was felt throughout the town. So attractive were the exercises that in 1910 local politician E. A. Sims proposed that private vessels be allowed to take part as scouts and dispatch boats, and he offered a $100 prize to any boat that could slip past the defenses at night or that could put a landing party ashore. His idea was not accepted and civilians remained as observers of the action rather than participants.[20]

For those who served the guns and mortars, target practice was serious business. It was their time to demonstrate what they had been trained to do, and to demonstrate how well they could do it. Their trade was different from that of other soldiers; they were the operators of a complex collection of machines and apparatus that waited for the enemy to come to them. When that time came, they would have to act with "smartness, coolness, self-reliance and discipline."[21] A battle with warships would be sharp and quick; there would be little opportunity for the coast artillery to become battle-hardened through repeated engagements. Therefore, their training had an urgency and importance not equaled in other services.

The life of the coast artilleryman was one of constant training, the goal of which was to learn to hit the target at any point within the range of his battery, at any time of the day or night, during any sort of weather. One officer called it an art principally of details, which he had to admit were tiresome in the extreme.[22] Gun crews went through their drills over and over again, loading and unloading the dummy ammunition. At the fire control stations, observers tracked vessels moving in and out of Admiralty Inlet, and the men in the plotting rooms repeatedly recorded the travel and predicted the future positions of steamers, barges, and Alaska-bound freighters.

The officers worked hard to keep the exercises realistic. Sometimes, the Navy's Pacific Squadron would cooperate by taking up a variety of attacking positions outside Admiralty Inlet, providing everyone in the fortifications an opportunity to prepare the defense based on the maneuvers of actual warships.[23] When Garland Whistler commanded the defenses, he arranged in 1910 a two-week stint at the batteries, hoping to provide as strenuous a routine as might be experienced in actual conflict. A series of mock naval attacks of several different kinds lasted throughout the final evening of the event. Whistler took pride in the response shown by the defense; after the bugler sounded a call to arms at 3:10 a.m., the entire command was in order two minutes later.[24]

The high point of the drills was the service target practice, when crews got to fire ten or more rounds at a target towed by one of the artillery district vessels. It was an intense and memorable experience as each man moved swiftly through his duties:

> It's great to see the big disappearing guns work. The men run out with a truck on which is a large projectile. At the command "Load!" they run the truck to the breech of the

Plate 6-3. Coast artillery service meant constant training, and here a crew goes through its paces at Battery Kingsbury, Fort Casey. The men on the long rammer staff are forcing the shell into the breech of the gun, and just behind them, two men are running up with the long bagged powder charge. Off to one side stands the gun commander, stop watch in hand. In the right foreground are two additional projectiles, ready for loading. The cylindrical projection on the front of the shell was an aid to piercing the heavy armor of warships. *Washington State Parks and Recreation Commission*

gun, two men with a rammer push the projectile off the truck and into the gun and at the command "Home ram!" they ram it up into the gun until it sticks. The two men bring up a long bag of powder and this is shoved in behind the projectile. The breech-block is swung into place and locked; the primer is inserted and the man yells "In battery—trip!" A man below jerks up a lever and the big gun rises, pokes its nose over the parapet. By this time the range setter calls

"Range set!," the gun pointer calls "Ready," the gun commander yells "No. 1—Fire!"—the gun pointer pulls the magneto lever, there is a blinding flash of flame then a deafening roar and jar—the gun settles back into its first position while the projectile goes hurtling out thru the air. Ten—fifteen—twenty—twenty-five seconds pass, and then way out in the straits, right up close to the target, a cloud of water rises—a hit! Five seconds later the piece again sticks its nose over the parapet and sends its steel messenger singing out over the water.[25]

The battery commanders strained to make the results of the target practice appreciable. In the early 1900s the sum of the practice was represented simply by the number of hits on and near the target as placed on a small outline of a ship's deck or profile; by 1913, the firing was judged through a complex mathematical formula with ten variables.[26]

Even the most painstaking management of statistics paled in comparison with the intent of the practice, and what every coast artilleryman longed for was the chance to fire at a real moving vessel. That never came to pass during the height of the defenses, although in 1920 some rounds were sent into the hulk of a battleship scuttled outside of Pensacola Harbor.[27] Not to be denied, an officer at Fort Worden created the next best alternative. Equipped with the data from a recent target practice and a set of drawings detailing the interior structure of a warship, he tracked methodically the hypothetical course of each projectile as it struck the vessel, noting the point at which it perforated the phantom armor and its would-be path as it crashed through nonexistent storerooms, crew spaces, cabins, and decks.[28]

Plate 6-4. A detachment at Fort Casey "filling and fusing" ten-inch projectiles. Although the magazines were large, there was seldom much ammunition kept on hand. Many shells were left empty, and when service practice came near, the empty shells were cleaned, painted, and as shown here, filled with an explosive charge and fitted with a detonating fuse in the base. The projectiles, gun carriages, electrical equipment, emplacements—in short the entire inventory of a coast fort—called for an almost continuous maintenance routine. *Washington State Parks and Recreation Commission*

The results of the practice were measurable and laudable. By 1906 the accuracy, rate of fire, and target range had so increased over the previous six years that the chief of artillery estimated that the best trained detachment was about sixteen times more efficient than the best trained detachment in 1900.[29] In Puget Sound, the excellent shooting of the 85th Company, manning the ten-inch guns of Battery Worth in 1908, elicited a favorable mention from the secretary of war.[30] In 1913 batteries at the three Admiralty Inlet forts ranked among the highest of all those participating in the nation.[31]

Trained and dependable men were fundamental to the success of a highly technical system although it seemed that awareness was slow to arrive. Writing in 1908, Luke E. Wright, the secretary of war during Roosevelt's second term, described a central dilemma. "Without an adequate force of trained men capable of handling all these different elements of defense our seacoast fortifications are useless and all the expenditures made upon them— now amounting in round numbers to about $85,000,000—are worse than wasted, for they have lured us into a false sense of security and protection."[32]

His point was well-taken. The proposals of the Endicott and Taft Boards had given no regard to the troops that would be required to serve the guns and mortars. Batteries were built and cannon mounted free of any connection with the men needed to turn them into weapons. A force of at least twenty-five thousand was necessary for the defenses as contemplated in 1895, a number then equal to the entire standing army.[33] While the number of officers and men assigned to the fortifications over the years never met the number required to man all the guns, the artillery service did grow expansively, a bloom forced by the national commitment to modern coast defense. It became the army's first branch exposed to an interrelated network of modern weapons, and achieved a distinctiveness peculiar to the operation of coast defense armament. Eventually, in recognition of its place apart, the coast artillery became a separate corps with an organization independent of the balance of the army. Its status helped create a touch of elitism among its members. Despite their self-esteem, the limited population of coast artillery units made it inevitable that they would have to share some of their special competence with someone else. That someone else was the state militia. So

important did the militia troops become that they not only had a place serving artillery pieces that would have otherwise gone unused, but also acquired other duties protecting the fortifications from land attack.

There was no coast artillery force to speak of prior to 1901. For many years, the Artillery had been grouped into regiments, an organization that emphasized the Artillery's role as a support for the cavalry and infantry that formed the bulk of the army's manpower, and many of the artillery units functioned as infantry in an assortment of scattered garrison posts. In 1887, few batteries had had any target practice within the preceding ten years, and others had not even fired any cannon.[34] The fundamentals of coast artillery had almost no standing and there was no consistent attempt to prepare a defense based upon the weapons available.

The rejuvenation of the coast defenses catapulted the Artillery out of its post–Civil War torpor and moved it toward the collection of skillful cannoneers whose responsibilities had little to do with the heritage of horse-drawn caissons and ponderous smoothbores. High-powered rifles on disappearing carriages or groups of heavy mortars could not be commanded properly by officers steeped in the ways of field artillery. The new coast artillery had to be guided by those with an understanding of

> military engineering, gun construction and the metallurgy of gun metal, interior and exterior ballistics, steam and mechanism, electricity and mines, chemistry and explosives, military science, telegraphy, photography and cordage in addition to being master of all that is contained in the 500 pages of the *Manual of Heavy Artillery Practice*, to which could be added a familiarity with the armaments of the world, naval tactics, attack and defense of places, siege operations, . . . organization and administration of fortress artillery, and many others which are artillery necessities.[35]

In short, the new coast artillery was highly complex, demanding of its officers and men a combination of intelligence and ability that was not as keenly felt in the other branches of the army.

Coast artillery materiel was understood as a collection of specialized tools and machines. Because of that view, the manning of the equipment achieved a special separateness not akin to the rest of the artillery service. Coast guns did not, could not, should not move; they were where they were because that was where they were needed and nowhere else. Moreover, the method of locating the targets, of supplying ammunition, of caring for the weapons, was totally different from field artillery practice. Throughout the 1890s, artillery officers clamored for an organization that would recognize the differences inherent in coast defense service.

In 1901 Congress passed a reorganization bill that answered some of the needs. All the existing artillery regiments became part of a new Artillery Corps headed by a general officer, which enhanced the standing of the new body among the other branches of the service. The bill also established two components within the Artillery Corps: coast artillery and field artillery. It called for about four times as many companies of coast artillery as it did batteries of field artillery, a recognition of the early dominance of coast defense in the new Corps.[36]

The field artillery retained its old battery and regimental designations, while 126 new numbered companies constituted the coast artillery units. The new companies formed rapidly between February and October of 1901. In the expansion many of the old regimental artillery batteries gained new classifications as coast defense troops. In Puget Sound, for example, Battery B of the 3rd Artillery Regiment occupying Fort Flagler became the 26th Company, Coast Artillery. Others were expansion units, formed by taking portion of the strength from a newly designated company to create the nucleus of another organization. Thus a few weeks after the 26th Company was born, some of its officers and men were parceled away to develop the 94th Company, Coast Artillery.[37]

The existence of the Artillery Corps helped nurture the image of coast artillery units as a unique class with duties significantly different from those of field artillery soldiers. It also fostered a cluster of talented officers dedicated to the improvement of coast

artillery, and convinced of its special attributes. The officers were a small group. Most had been classmates at West Point, and if they did not know each other personally, then they were at least familiar with names and abilities. The first years of the twentieth century brought among these men an intellectual ferment that produced a host of valuable ideas. It was a rich and exciting time, a period when the systems of manning the weapons and locating targets received the most careful attention. One witness later recalled

> [Clint C.] Hearn, with projects and inventions sticking out in every direction, like Jupiter's thunderbolts, making him dangerous to tackle, coordinated with [Garland] Whistler and between them was produced the present plotting board, [Isaac N.] Lewis was after the D. P. F. [depression position finder], and Whistler, never still, was stirring things up at Fort Wadsworth and giving night exhibitions to the Chief of Artillery of call to arms with the suddenly awakened men clad in earnestness and zeal, but little else… The revolution was completed, the artillery was awake, the woods were full of live working artillerymen and one could scarcely through [sic] a stone without hitting some worker, so absorbed on his problem that he would neither have noted the coming of the blow, nor cared if he was hit.[38]

There was notable disparity between the fervor of the coast artillery and the more mundane routine of the field artillery. A partner to that disparity was the widely held opinion, in which the chief of artillery himself concurred, that the two types of artillery did not belong together in any fashion. The needs and duties were too dissimilar to be considered under the same major division of the army. There was no tactical relation between the two and the secretary of war believed them to be fundamentally different, just as cavalry was from infantry.[39]

Congress passed legislation in 1907 that recognized the differences. The coast artillery was made into a corps of its own and independent of the mobile armies of the United States. Its sole function was to serve in the harbor defense. The act that created the Coast Artillery Corps also addressed some other shortcomings, particularly the size of the company organization. Each of the coast artillery companies formed following the 1901 act had the same number of men, even though that number was not always a good fit for the gun battery to which the company was assigned. To man a battery of two twelve-inch guns, ninety-four individuals were necessary, but only eighty-six were needed for two ten-inch guns and sixty-three for the same number of six-inch cannon.[40] With a fixed company strength, some units would have more than enough men while others would have too few. The 1907 act made the strength of the company vary in accord with the number and caliber of weapons to which it was assigned. It also provided for new grades and increased pay for specially trained enlisted men, and added forty-four more companies to the Corps. The new units were mine companies, organized specifically to support the submarine defense, a feature overlooked in the general reorganization of 1901. Two of them, the 149th and the 150th Companies, were stationed in the Coast Defenses of Puget Sound.[41]

The Coast Artillery Corps was something of a phenomenon. Its inception had taken place over a very short period of years and more remarkable, it had no organizational relationship with the rest of the army. It was an independent force composed of equally independent companies. Instead of being established along traditional military lines, the Corps was a conglomerate of units assigned to specific geographic locations. The harbors themselves provided the basis for command and administration. For the companies stationed there, the harbor was the place of their peacetime garrison as well as their tactical position in a future war. The strong regional flavor was evident in the 1916 renumbering of the coast artillery companies; in that year the organizations gave up their old designations for new ones that linked them directly to the posts in which they served. Thus the 30th Company, stationed at Fort Worden, became the 1st Company, Fort Worden; the 62nd Company, also at Fort Worden, became the 2nd Company, Fort

Worden; and so on throughout all the forts. The appearance was altered during the World War when the units were redesignated again, this time using the emphasis of the regional coast defense rather than the separate post. The 1st Company, Fort Worden, became the 1st Company, Puget Sound; the 2nd Company, Fort Worden, became the 2nd Company, Puget Sound; and the process was repeated for all the active units.[42]

There were basic problems that reorganization could not resolve. Desertion, low reenlistment, and inadequate strength hampered the business of coast defense. Improvements in the lot

Plate 6-5. Peeling spuds for dinner. Each of the coastal forts operated as a small community and there was always plenty to do. One soldier described it in a letter home in 1918, noting that his company had to "to stand guard duty, post fatigue [a general work party], kitchen police and drill. I was on guard duty from Sunday night to Monday night. Today I get artillery instruction and tomorrow I will probably be on K.P. I like the artillery dope but I don't like the work very well." *Jefferson County Historical Society*

of the soldier diminished the first flaws, but the number of men on hand to man the defenses was always insufficient.

Desertion was a common problem in coast artillery units. It rose and fell in conjunction with a number of factors, living conditions being paramount among them. The construction of temporary barracks behind the batteries at the Admiralty Inlet forts did away with the canvas shelters, and the replacement of the temporary structures in 1904 by the heavy frame buildings fulfilled the army's intent to make its permanent reservations healthful and attractive homes.[43] The dismal isolation of Fort Flagler and Fort Ward, however, was profound. Desertion rates at the two forts remained high, and in 1912 were among the highest in the Army.[44] The remoteness of Fort Flagler, as well as its inferior position within the arrangement of the defenses, inspired the move of the coast defense headquarters from there to Fort Worden in 1904, and certainly the eager promotion of the idea by the Port Townsend Chamber of Commerce and the state's congressional delegation must have helped as well.[45] The addition of recreation buildings at all the posts was an attempt to improve the atmosphere in the garrisons, and probably the intense promotion of an inter-fort baseball competition also had its roots in the wish to reduce desertions.

Given the circumstance prevailing in the Puget Sound defenses in the early 1900s, it was easy to grasp why a man did not reenlist if he at least elected not to desert. The reasons for one were certainly related to the other. Reenlistment was essential for a technical service because only long training could develop the kind of individual wanted in the coast artillery. Yet as they reached the end of their three-year enlistment, the carefully schooled electricians, telegraph operators, engineers, and other specialists departed, leaving new men to be trained each year. A basic disincentive was that the men were paid the same as those in the infantry or cavalry, even though the mechanical skills which the army sought were well paid in civilian life. It was difficult to persuade a private being paid thirteen dollars a month in 1903 to reenlist when civilians performing comparable work could expect from

two to four dollars a day.[46] The enlisted soldier's finances improved in 1907 when Congress increased pay for those with special training and provided additional pay for length of service, changes which were enhanced in 1909 by further increases.[47] Reenlistment and enlistment improved thereafter.

Desertion and reenlistment influenced in small ways the day-to-day population of each coast artillery company, and the capability of each unit in the defense fluctuated accordingly. There were three more fundamental components of strength: the number of men required to man all the defenses as built, the number of men authorized by Congress to participate in the defenses, and the number of men actually enlisted in the service. All three changed as more guns were mounted, or as the planned strength of the army altered, or with the varying ability of army recruiters to attract men to coast artillery companies. The key to a sound coast defense was a sufficiency of well-trained men, yet neither the authorized nor actual strength of the coast artillery was ever adequate to supply crews for all the weapons. How the coast defenses maintained a balance between the numbers of men they needed and the numbers of men they actually had was a story of continuing compromise.

The men required for an adequate coast defense was a fixed quantity based upon the catalog of guns, mortars, and other appliances in the fortifications. The figure was precise. Any amount less and the defenses would be served inadequately, any amount more would be a waste of manpower. Unlike the cavalry or infantry, the coast artillery did not become more menacing as it acquired more personnel. What mattered most was the weapons in place and how many of them had complete manning details.

The realization came early that there were far more defenses to occupy than there were troops on hand. The 1901 act fixed the coast artillery at about 13,700 men, an increase over the previous 8,000 men for all the artillery, although the amount was far short of the 23,000 necessary for the armament in existence in 1903.[48] By the next year, the number authorized was not enough even to man one half of the guns and mortars already mounted, and more

cannon were being emplaced every month.[49] In 1906, the chief
of artillery estimated that he would need 41,800 men to provide
a single relief for all of the defenses. There was some response in
1907 when the authorized strength of the new Coast Artillery
Corps rose to its pre–World War I peak of 19,300.[50] If the defense
contemplated by the Taft Board were added to the existing totals,
there would have to be a Corps of more than 47,700 men.[51]

Congress was unwilling to consider a force so large. By 1908,
the Coast Artillery Corps already constituted fully a quarter of
the army, and most legislators were unimpressed with the insistence of the chief of coast artillery that the defense required an
organization of the size being discussed. There was a common
belief that should danger threaten, gun crews could spring from
the American heartland in the tradition of the minuteman who
stood forever ready to take down his long rifle and powder horn
to answer national distress. It was a confident image founded in a
mixture of history and mythology. Those charged with the coast
defenses paid it no heed; they were sure that a future war would
be swift and without sympathy for credit given to native folklore.
When war came, it would appear like an "avalanche upon the
unprepared and overwhelm them in its course"; to propose that
volunteers assist in the defenses, one officer grumbled, was "the
drivel of a maundering idiot."[52] All coast artillery professionals
agreed that the troops they needed could not be improvised. With
not even a dim hope of receiving enough men to fill the batteries,
they turned to another source: the state militia.

The idea of using the state militia appeared while the first batteries were being built, and a little later, in 1896, members of the
regular artillery helped train part of the Massachusetts Volunteer
Infantry in the ways of coast defense.[53] The Spanish-American
War spurred the interest of Atlantic coast states and, perhaps
encouraged by the budding enthusiasm, the secretary of war
announced a policy in 1902 of promoting the affiliation between
the state troops and the regular coast artillery units. Fortification
service by the coastal states' military was ideal militia work, he
said, since it was always to repel invasions and the duty was less

onerous because it was close to business and homes and did not
require distant service.[54] Moreover, the civilian population of the
coastal areas contained electricians, machinists, and engineers, the
best possible material for coast artillery. And, as one artilleryman
had noted earlier, it could do no harm to become closely allied in
work, feelings, and interests with such a large body of voters as the
seaboard militia might become.[55]

The federal government urged the states to organize their
forces along the lines of the regular army and provided equipment
and training as inducements. The State of Washington was slow to
form its units. Early interest in 1903 was thwarted by the lack of
any mention of heavy artillery in the law that established the state
militia.[56] In the absence of legal authority, the state agreed in 1907
to participate by way of experiment in joint army-militia coast
defenses exercises. Eleven companies of state infantry went to the
gun batteries to see how much they could absorb during the brief
training period. Many of the men were from eastern Washington
and had seen neither salt water nor coast defense cannon before
their sojourn to Puget Sound.[57] For ten days they wrestled with
the strange equipment and then culminated their encampment by
firing service rounds. The experiences and practices locally and in
many coastal states demonstrated that the troops could learn the
unfamiliar quickly enough to be considered an effective supplement to the regular coast artillery units.

In 1907 the War Department reorganized the manning of the
defenses and made the militia of each coastal state an essential
partner. The militia and the army now shared the gun defense; one
half of the crews were to come from the militia and the other half
from the regulars. The plan called for the state troops and the federal troops to hold joint training exercises each year. The quality
of the exercises was important, for it was during the brief annual
service that the militia would receive its most valuable knowledge.
To ensure that the training would be thorough, each militia unit
was linked to a regular coast artillery company, and each man in
the unit was linked to a man in the regular company. The regular company would instruct the militia company in its duties,

and each man in the regular company would instruct his militia partner in his individual responsibilities. Once fully schooled, the militia company would be assigned to a single battery on a permanent basis so that in peace time it could become familiar with the battery's every detail and the water area covered by its guns.[58]

The regular troops still carried the largest part. They alone were responsible for the defenses in Hawaii, the Philippines, and the Panama Canal, as well as all the searchlights, power plants, and mines. Only in the United States did they divide the armament with the militia. The apportionment was still critical. If the militia could produce troops in the numbers then estimated as necessary—about 18,500 men—then the existing strength of the regular coast artillery was closer to what was required under the revised manning program and the nation was nearer to an adequate coast defense.[59]

The fortifications could be maintained and the men trained effectively if three conditions prevailed. The first was the concentration of the available companies in each coast defense to provide as complete a manning detail as possible at one post, leaving caretaking detachments at other installations. The technique, which had little effect in Puget Sound, allowed more intensive fire control practice and a better standard of discipline. The second condition was that the number of officers and men necessary for the searchlights, power plants, and mines be maintained in full. The final condition required that the number of men for one-half the continental gun defense be the basis of the authorized peace time strength of the coast artillery.[60]

That the chief of artillery could consider a greatly reduced strength was due to the participation of the militia; the states, sensing their importance and enjoying the sudden attention, turned to nurture their companies. By 1909, there were 138 coast artillery militia units, formed or in the making, scattered among the coastal states.[61] Washington had amended its militia law and had converted five of its infantry companies in the same year to coast defense duties, and it hoped to expand to a corps of eleven companies by the start of summer training.[62] In the general

enthusiasm, the state of Ohio came forward and offered to form several artillery companies for use in the Atlantic coast forts.[63]

The states had a formidable task. The strength of any militia unit was of course a function of the population of the state, but the fortifications had been built with thought given only to the needs of defense. In some areas the number of gun and mortar batteries far exceeded the potential of the state to supply its portion of the troops to man them. The Washington militia found itself precisely in that dilemma. Because the geography of Puget Sound required extensive defenses, the state militia of Washington would have to find some 1,900 officers and men organized into twelve companies, the largest quota in the nation.[64] The state was never able to expand beyond the five companies it developed originally before World War I. Equally limiting was the strength of those companies that were organized. The total of the troops in the Washington Coast Artillery Corps rarely exceeded three hundred men, which meant that every company was far below the number required to serve a battery. During the summer camp of 1912, two of the companies had fewer than thirty men and the average strength was only forty-four.[65] There could be no intimate connection between regular and militia units, and there could be no brotherly instruction of the state soldier by his federal counterpart if the strength of the units did not in some way approach equivalence.

The dependence upon the state militia had proved a great disappointment. As a national force, it had produced only a fraction of the men anticipated. While the more populous states, or those with small coast defense establishments, had little trouble filling their units, others had before them what seemed an impossible task of recruitment. Enthusiasm flagged, and in Washington two companies had to be disbanded because of their low strength.[66] By 1915, ten of the twenty-one coastal states had dropped their coast artillery militia.[67]

The regular army had somewhat better success. The scheme to concentrate the companies had been carried out and it seemed to be working. The War Department kept the mine companies at strength, apparently believing that the work was too complex to be mastered by the occasional participation of the militia, and there seemed to be enough men to maintain the electrical plants in operating condition. The other part of the 1907 program fared less well. The authorized strength of the coast artillery was not made equal to that necessary to man half the guns and mortars, and the actual strength was considerably less than that authorized.

Coast Artillery units seemed perpetually under strength. It had been almost impossible to fill each company to its authorized number of men during the rapid expansion of the artillery in 1901, and many units never did catch up. Even when the Corps could expect to share the continental defenses with the militia, it still had to prepare a force of more than thirty thousand officers and men to cover the fortifications in the insular possessions and the Panama Canal, the mines and other auxiliaries, and its part of the guns and mortars at home. It was easy to demonstrate the need; it was more difficult to interest Congress in expansion. Moreover, the actual strength almost never met the authorized strength, staying between seventeen and eighteen thousand men before World War I.[68] The Coast Defenses of Puget Sound fared reasonably well during the same period. Its complement of thirteen companies was among the largest of the garrisons, with an enlisted population of about 1,300.

Forts not on the continent were kept at full strength, a policy which reduced the troops available for use in the batteries guarding the harbors of the United States. By 1915 the deficiency in militia troops had grown, and the authorized strength of the Coast Artillery Corps was 48 percent of what was necessary to meet its already limited obligations in the home defenses. The lack of men meant that one third of the armament in the fortifications could not be served by either the states or the federal government.[69] Help came in the form of the National Defense Act of 1916, which called for a large increase in the army as a whole. Under the act, the Coast Artillery Corps was to expand over a five year period to 21,400 men, at last adequate to meet all its obligations. The states were still to be called upon to provide one half of the men required, and measures were planned to arouse greater interest within the

states. The arrival of World War I and the mobilization of all available manpower suspended the National Defense Act proposals and they were never implemented completely.[70]

War volunteers and draftees swelled the companies of the Coast Artillery Corps and the militia. As if to symbolize their cooperative role, the units of the regular army and the militia were renumbered. Following the formation of three additional companies in 1917, the regular units in Puget Sound became the 1st through the 16th Companies, and the militia troops, now recruited to a total of twelve companies, became the 17th through the 28th Companies, Puget Sound. An infusion of regular troops created the 29th through the 41st Companies, bringing the strength of the Puget Sound defenses to its high point of 4,700 men.[71] Regular and militia companies alike contributed to the seven artillery regiments formed from Puget Sound coast defense soldiers for service overseas, including the 63rd Artillery, the only regiment organized exclusively in the Coast Defenses of Puget Sound.[72]

After the war, the militia companies were disbanded and the regular troops returned to their prewar duties in their old stations, now substantially depleted of their armament. There was a minimal movement following the war to reconstitute the coast artillery militia. In 1921, the State of Washington organized several companies with a combined strength of less than two hundred men, beginning again a limited involvement with the coast defenses, which would last through World War II.[73] Of the regular Coast Artillery Corps in Puget Sound, nine companies were demobilized or inactivated within four years of the Armistice, evidence of the readjustments brought about by the reductions in the postwar military and the gradual demise of coast defense in general. Only four of the original thirteen companies remained to be absorbed into the regimental organization imposed on the Coast Artillery in 1924.[74]

The militia figured prominently in another aspect of coast defense: the protection of the fortifications from landing parties, a role known commonly as land defense. The designs for individual batteries or for complete fortifications usually did not take into account any considerations for the attack of the defenses by other than warships. There had been the same slight concern in Puget Sound. John Millis gave some passing thought to locating Battery Walker hard on the left flank of the Fort Worden mortar battery because the position was good for land defense, but he dismissed the idea as inappropriate.[75]

The reasons for what would be called the "peculiar obliquity of foresight" were not difficult to understand.[76] The coast defenses were founded on a single purpose and their only function was to halt naval vessels. There was little reason to believe that any force more powerful than a raiding party could be landed beyond the range of guns, and there was nothing to be feared from that threat since the fundamental assumption was that all of the area surrounding the permanent defenses would be in the control of the mobile army. In the 1890s land defense was little more than a good topic for discussion, and the conversation trended two ways. There were those who felt that, since to run past a coastal fort would almost certainly result in the destruction of the vessel, groups of infantry would have to be landed to capture the batteries and silence them. Others, considering the physical prohibitions in moving men and equipment in small boats and then assembling them into some sort of effective military contingent, believed that landings would remain useful only against "unprepared, disorganized and barbarous people."[77]

As the initial rapture of the new fortification program faded, many began to realize that the issue of land defense was very real. If war came, the mobile army could not provide enough troops for the protection of the fixed defenses in all the harbors, and without infantry and cavalry available to travel the countryside, the forts invited attack from small landing parties. There were darker thoughts. In the absence of a major mobile force, the defenses could be thwarted by an enemy fleet landing troops on an uncontested shore distant from the fortifications. From there, the invader could march upon an improved harbor, gain possession of the dock and terminal facilities, and begin an invasion in earnest. Once enough men had been landed, they could move to the

defenses which had been built to protect the harbor now occupied, capture the batteries from the rear, and perhaps even turn the guns on nearby cities.

A bizarre hypothesis was put together in Puget Sound. An "oriental nation," most certainly Japan, would land a strong body somewhere to the west of Fort Worden and attack Port Townsend. Taking the city, occupying Discovery Bay, and manning the guns of Fort Worden and Fort Flagler, they would then begin the creation of a formidable naval station and commercial port. With the two forts controlling Puget Sound and the city defended by strong field works, the isolated enclave would become a rallying point for a heavy immigration to the Pacific coast and would dominate a new Asian-Alaska trade route. Peculiarity of imagination aside, the same 1908 report identified the first Japanese objectives in a Pacific war as the Philippines and Hawaii.[78]

There were more realistic assessments of land defense problems in Puget Sound. Commodious Discovery Bay was undefended and was a very likely spot for a landing, Also, if a fleet bent on destroying the Port Orchard Naval Station managed to pass by the Admiralty Inlet forts intact, there would be no reason to face the guns and minefields at Fort Ward. Once in Puget Sound, ships could enter Hood Canal and put men ashore within easy march of the dry-dock and refitting yards, and never come close to the defense at Rich's Passage.[79]

The importance of the omission at last recognized, the chief of engineers directed in 1903 that land defense plans be made for the principal harbors, and included Puget Sound among the first to be considered. Maps were drawn up showing all the trenches, blockhouses, redoubts, and wire entanglements needed to protect the fortifications. The details identified all the prominent terrain features in the vicinity of the forts, and in the case of Fort Flagler, included photographs depicting the character of the beach and shoreline. The maps often sketched in earthworks on private land. Congress did not allow the purchase of property for the exclusive needs of land defense, a policy based on the presumption that there would be enough time after the outbreak of war—when land could be condemned in the name of national defense—to build the field works with civilian labor. Since most of the land defenses were planned for sites outside the reservations established for the fortifications, the Puget Sound forts showed little physical change when the land defense plans were approved, although a few rifle pits were dug near remote searchlight positions.[80]

It was one thing to dig trenches, it was another matter entirely to fill them with troops. Who was to wait for the enemy should he decide to come by land? Not the mobile units of the regular army; they would have a greater purpose than to be spread thinly among the defended harbors. Not the men of the Coast Artillery Corps; their duties were at the guns and to take them away would weaken the coast defenses. The answer was the same as it had been when the army had sought more men to serve the gun and mortar batteries: the state militia.

The 1907 policy that asked the coastal states to supply half of the men necessary for the fixed defenses also placed upon the local militia the defense of the forts themselves.[81] The militia assigned to the batteries was known as the Coast Artillery Reserves, and militia units to be charged with land defense were known as the Coast Artillery Supports. The Coast Artillery Supports were mobile forces organized exclusively for the protection of the fortifications, and they were to have been given a distinctive insignia to show at a glance the special nature of their duties.

There was another classification of mobile troops, the Coast Guard (not to be confused with the United States Coast Guard), which was to have been organized as a field army operating beyond the influence of the fortifications with the general mission of preventing a land assault upon cities, arsenals, shipyards, or other critical points. Of the two organizations the Coast Artillery Supports received the most attention since they would have carried the brunt of any landing parties bent upon the destruction of the defenses. However, the crucial need to find enough men for the Coast Artillery Reserves damped the formation of special land defense troops, and the dismal maw of the militia coast artillery quota swept away any belief that the states could provide the

Coast Artillery Supports as well. The Coast Guard continued as a desirable idea, yet its realization remained even more unlikely than that of the Coast Artillery Supports.[82]

The dilemma of land defense emphasized that the collection of concrete fortifications were artillery positions only, without any of the characteristics of a fortress. The defenses were a tool to defeat warships, and to avoid possible damage during an engagement, the Corps of Engineers gave a paramount place to dispersion. As soon as they opted for dispersion, they gave up the opportunity for an economical land defense. Each battery could have been designed with some sort of landward face that would have converted it from simply a seacoast artillery battery to a redoubt, yet the question of manning the installation would be unanswered. The coast artillery units, whether militia or regular, could either stand by their heavy ordnance or equip themselves with rifles and prepare for an assault from the rear, but they could not do both. Confining the troops in specially built structures had an additional disadvantage in that it implied that an enemy would be allowed to come within rifle shot before being countered, although any serious landing party would bring with it light artillery pieces that could bombard an emplacement from several thousand yards away. Given the conditions, the best land defense was one that depended upon mobility rather than the fortification science practiced for seaward protection.

For the engineers who devised the first land defense plans, the whole matter was of little interest. They were far more committed to, and later criticized for, an exclusive devotion to what they called "the technical requirements of the defense," or how many seacoast guns of what type were necessary and where.[83] Land defense was not important when the appropriations were for items narrowly related to coast defense alone. Mulling over the vulnerability of some of the more far flung secondary and supplementary position finding stations, John Millis shrugged it off with the comment that their occupants would have to do the best they could with temporary field works, and would just "have to take their chances at least until the condition of the national treasury may be improved."[84]

One of the best aids for land defense had been installed already: the heavy cannon themselves. Almost all, however, had been mounted in emplacements that restricted their fire to the front. The mortars were an exception. The cannon could be traversed in a complete circle and with a little maneuvering, it was possible that the seacoast mortar could be used as a devastating weapon against shoreward targets. Its high trajectory had the additional advantage of being able to soar above any hills that lay between the battery and target, a benefit not found in the guns. The test firings into Port Discovery from Fort Worden indicated that the impact of the shells could be directed from a distant point, and a few years later in another coast defense, a mortar battery fired upon an infantry trench as an experiment.[85] While these efforts indicated the flexibility of the mortar, no fuse or projectile suitable for land targets was developed, and it remained only an interesting possibility. The construction of emplacements that would permit the land defense use of coast guns in general did not become policy until 1915.[86]

The efficiency of the fortifications in Puget Sound was defined in large measure by the personnel assigned to them. An ambitious building program had outstripped the desire of Congress to support a large standing force in the coast defense and the War Department had been unable to persuade the coastal states to take up more than a token share. That the states might be the source of the necessary manpower was a delusion, probably engendered through the accumulation of statistics about the "males available," extracted from state population figures. Such numbers were large enough to hold forth the hope that there might be a willing body of men ready to become gun crews in the fortifications, not to mention the infantry necessary to protect the works from land attack. However, the demands of defense and the population of the states were not balanced; in the case of Washington, the military needs were far in excess of the few militia troops. Despite those limitations, there arose a well-defined sense of coast artillery service as special and requiring the best men that the Army had. The Coast Artillery Corps was one of the "highbrow branches of the service," as one old soldier recalled, a collection of elite troops manning unique weapons.[87] World War I brought an end to that appreciation as it signaled the demise of the spirit and form of fixed defense as an important element in the protection of the United States.

To Cope with New Conditions

"We of the coast defense must sit behind more or less antiquated weapons. It cannot be denied that our present coast armament is not sufficiently powerful to cope with the new conditions with which we are faced."
—*Anonymous, 1915*

I t was the biggest party that Port Townsend had ever seen. They closed the downtown streets to make room for the two thousand Puget Sound artillerymen that crowded in for the celebration. A band played right in the middle of it all, and there was dancing ("All Dancing Women Wanted," the local paper had advertised a few days before) and fireworks and happy tumult decorated by clouds of confetti. The din was so great that the speakers could not be heard but no one seemed to care as the good time continued on into the small hours of the morning. And why not? The Great War was over.[1]

And it would soon become apparent that the heyday of coast defense was over as well. World War I ruptured the careful pattern of the previous decades. During the war years, the fortifications and the Coast Artillery Corps began a rapid, irreversible decline. By the early 1920s, the evidence was everywhere. Many batteries stood empty of their guns, and garrison buildings decayed for the lack of troops to occupy or care for them. Entire posts lay abandoned. What was less apparent was that while there were sound technological reasons for reevaluating the fortifications after the war, the speed with which the fixed defenses were discredited was a phenomenon cultivated by the attitude of the professional artillery soldiers themselves.

The inability of Endicott and Taft period works to keep pace with continuing developments in naval engineering hastened the passing of the coast defenses. The fortifications had been designed to combat armored ships at a time when the vessels were in the first stages of development. The coast defenses were made competent to attack a certain type of target, and there had been no anticipation that the target would change so rapidly. Without a policy of restructuring the defenses to meet new capabilities afloat, they were soon surpassed.

Armament provided the paramount example. The barbette and disappearing carriages installed before the war permitted just a small elevation for the guns mounted upon them, only about twelve degrees.[2] It was a significant point because the greater the elevation, the greater the range, and in that way the carriages actually acted to restrict the potential of the weapons mounted upon them. The ordnance designers of the 1890s believed that there was little need for greater ranges because the range of naval guns was even less than that of the shore armament. The design of seacoast carriages was founded on the faulty premise that the mechanical complexities of devising a shipboard mount would prohibit long-distance firing from naval vessels.[3]

Yet it was not long before naval guns could elevate to angles previously thought impossible and ships gained the ability to strike at distances beyond the reach of the defenses. There was still some comfort in the more accurate fire control and communications systems of the coast artillery. While ships might be able to open fire at great distances, so went the thinking, their poor position finding would not allow those far-traveling rounds to have much effect. Once within the range of the defenses, the vessels would be decimated by the telling fire of the coast guns.

The confidence was short lived. In 1911, the navy conducted gunnery tests that demonstrated in chilling fashion that any modern navy outclassed the coast defenses. A fleet of battleships, steaming at a modest six knots in Chesapeake Bay, fired upon the *San Marcos* (the former *Texas*, renamed for the occasion), and wrecked the target ship completely. Army officers observing the practice watched in awe as salvo after salvo tore into the ship from up to twelve thousand yards away.[4]

It was a staggering lesson for the coast artillery. First, the range was in excess of what was then considered workable for the fixed armament. In the same year as the *San Marcos* firings, the average range for heavy shore gun practice had been a little over eight thousand yards; the coast artillery troops were practicing at distances unrealistically short.[5] The range at which the Navy gunners had so effortlessly struck the *San Marcos* were the utter limit for most of the guns in the defenses. Second, the accuracy of the shelling indicated that naval position finding equipment was at least equal, and possibly superior, to the horizontal base system; certainly it was far better than thought. The shipboard optical devices enabled the precision firing upon a target much smaller than the collection of batteries presented by most coastal fortifications. Moreover, the practical limit of much of the coast defense position finding equipment was twelve thousand yards.[6] Even if the guns had ranges equal to those of naval ordnance, there was no reliable way to locate targets at great distances. They would have no help from the fire control network that was supposed to be the core of effective coast defense.

The gunnery exercise pointed up another weakness, this time in the form of the emplacements. The batteries had been designed when naval gun fire was restricted to flat, almost horizontal, trajectories. Shells directed at the fortifications would tend to pass over them and fly harmlessly to the rear. Disappearing gun batteries in particular reflected a design that presumed little threat from gun fire arriving at more than slight angle, and all batteries were

therefore devoid of overhead protection. However, the shells striking the *San Marcos* plunged down steeply and the relative security of manning coast defense guns, as well as their comparative invulnerability, became questionable.[7]

Not long after the *San Marcos* practice, modern naval ordnance increased its range to eighteen thousand yards—and advanced to twenty-five thousand yards by 1915—and the guns were mounted on faster ships with tougher armor. Against such opponents, the defenses could do very little. Many agreed with the artillery officer who said in 1912 that "it may be frankly stated that the sinking of such a ship by long range fire of coast guns is entirely out of the question."[8]

The originators of the fortifications had immersed themselves in contemporary naval technology to produce specially tailored defenses which, for a brief time, more than equaled the warships. However, in their concentration, the designers neglected the most common method of keeping a first-rate naval force up to date. A good navy was never static; the fleet was always changing, discarding vessels made obsolete by new ideas and introducing others that embodied concepts or apparatus considered more desirable. One of the best examples was close at hand, the navy of the United States. Between 1899 and 1916, the number of battleships and cruisers doubled from thirty-six to seventy-seven, the result of incremental annual additions to the fleet.[9] There was no similar process for coast defense. There were those, chiefly members of Congress, who assumed that the batteries, once built, would be useful for an indeterminate time. The assumption was an unmurmured assent to the long-standing, however erroneous, belief that money spent on fortifications would not have to be spent again, or at least not until some point well into the future. That form of economy might have been valid at a time when innovations in military science were few and the rate of their emergence slow, but it was dangerously out of fashion by the date of the Endicott Board. Under the influence of deceptive permanence, the War Department for many years estimated the adequacy of the nation's defenses by comparing the batteries completed against the total recommended by the Endicott Board and, later, the Taft Board. Meanwhile, naval vessels continued to change, and it was not long before fortifications were still being built to defend against warships that were no longer a part of any main battle fleet. It was the lack of a policy of continuing fortification development that, more than any other reason, created the conditions for the demise of fixed coast defense.

An early episode of World War I demonstrated the ultimate hazard of depending upon defenses surpassed by the weapons which could be brought against them. The collapse of the respected Belgian fortifications in 1914 forced the realization that America's harbor defenses were critically dated. Brigadier General Dan C. Kingman, chief of engineers, suggested a new approach to fortification construction, one that abandoned old emplacements and weapons as they became no longer useful, and that planned for new defenses to take their place. A constant rate of replacement, about ten per cent of the emplacements annually, would allow the defenses to remain minimally current. He conceded that there had been modernization—the installation of ammunition hoists and the widening of loading platforms—but that was tinkering, analogous to a ship's overhaul which, although necessary, did not prevent its obsolescence. He emphasized that the art of fortification was a progressive one. "However carefully planned and constructed," he said, "a battery must always pertain to the date when completed and must be out of date in so far as it relates to things which have been discovered or developed since the battery was planned." Permanent defenses were not permanently useful, Kingman repeated. "A fixed project for seacoast defenses can never be adequate, and its obsolescence must begin before it can be completed."[10] It was all too true and about thirty years too late. The rapid approach of American involvement in the war meant that there would be no opportunity to start a new fortification program based upon such a pragmatic philosophy.

✳ ✳ ✳

No navy threatened the harbors of the United States in 1917. The coast defenses, both the weapons they contained and the troops stationed in them, were not necessary for national defense in their original role and more important uses were found for them. During the war, almost four hundred guns and mortars were removed from their emplacements so that they could be remounted on wheeled carriages or tracked carriers for service overseas, or shifted to railway carriages.[11] It was not unprecedented; the idea of borrowing the guns from the fortifications for a purpose other than coast defenses dated at least to 1906, but the thought had been that the guns would be returned.[12] As it transpired, stripping the batteries in World War I was for the most part a one-way operation.

The practice depleted the defenses of Puget Sound. All of the five-inch guns were taken away, and all of the six-inch as well, save for Battery Harrison and a pair retained at Battery Tolles. The eight-inch guns at Fort Ward went, barged to Seattle and then moved out at night by rail to the east coast.[13] The ten-inch guns of Battery Kingsbury, Battery Rawlins, and Battery Randol were dismounted and all the other ten-inch guns were listed for removal.[14] At the mortar batteries, the two forward cannon in each pit were also dismounted, destined for railway carriages. Only Battery Schenck retained its full complement of eight cannon. The mortars remained in their emplacements at Fort Casey pending the completion of a new mortar battery at Westport, protecting previously undefended Grays Harbor, on the Pacific Ocean coast about ninety miles southwest of Seattle.

The diversion of some armament away from the existing fortifications to new positions was a policy introduced by the Board of Review in 1915, and intended to protect previously undefended harbors that could be considered potential landing sites.[15] After the simple foundations were ready at Westport, the mortars stayed at Fort Casey, although the five-inch guns of Battery Lee were taken to the same vicinity and mounted on simple concrete blocks.[16] Similarly, two six-inch guns were brought from Fort Stevens to Willapa Bay, south of Grays Harbor, and emplaced there. It was those weapons that would be transferred to Battery Tolles in the 1930s.[17]

A few guns were to have been mounted on unarmed transport ships, and the five-inch guns of Battery Warner were marked for the *Dix*. The *Dix* was too busy to be shunted to Puget Sound to receive the cannon, and they lay on the dock at Fort Ward for a year and a half. In 1919 the guns were remounted in their emplacements, the only cannon in the defenses that were returned once they had been removed.[18] By war's end half of the cannon in Puget Sound were gone.

Mobile artillery was more than a wartime expedient, and the coast artillery was quick to seize upon the idea for seaward defense. Initially, the idea embraced only twelve-inch guns on railway mounts, then expanded rapidly to include all sizes of cannon, and soon grew to a point where many supporters advocated the use of mobile artillery to the exclusion of all fixed weapons. The mania for mobility expressed by an increasing number of coast artillerymen helped limit the development of permanent fortifications after the war.

Coast artillery troops had gone overseas and instead of living out in combat the drills they had practiced with disappearing guns and seacoast mortars in peacetime, they manned heavy field artillery, trench mortars, and anti-aircraft guns. Their greatest utility in warfare had little to do with thwarting a naval attack on an American port city. It was as if the whole idea had been wrong to start with.

There had often been suggestions that the coast artillery might serve in other roles, maybe even overseas, if there were no threat of naval attack. By 1910, the War Department had come to view the Coast Artillery Corps as having become reasonably proficient in its primary duties and, not willing to allow such a large body of men to remain skillful in one task only, cast about for alternative uses.[19] Infantry tactics seemed the best study. All coast artillery units received one month's training in infantry work, and the Corps provided a separate brigade of infantry towards the Maneuver Division, a gathering of the Army's scattered troop

units in 1911.[20] Periodic infantry training continued until the outbreak of war.

Siege artillery also seemed a natural co-function of coast artillery, and the chief of staff urged the training upon the troops. Erasmus Weaver, then chief of coast artillery, acceded to the direction and organized a siege battalion. Weaver did not care for the idea. He had not enough men in the Corps for a single manning detail, let alone provide for a separate organization dedicated to siege work. Also, and perhaps more parallel with Weaver's thinking, man-handling the clumsy siege pieces was a job more befitting the field artillery. In the summer of 1913 he succeeded in having the siege duties recognized as a function of the field artillery.[21] Despite the change, practice in field artillery continued as part of land defense preparedness, and almost every regular coast artillery company in Puget Sound had assigned to it a variety of horse-drawn field guns.[22]

Given the diversity of background in other types of service, the war-time employment of the Coast Artillery Corps was not unusual. More worthy of note was the ease with which the Corps took to a wide assortment of duties and equipment. Troops manned British and French weapons as well as their own wheel-mounted former coast defense cannon, and they quickly learned new ways of preparing target data. Additionally, the overseas forces were organized not as the independent companies of coast defense, but as regiments, which were easily adapted into the existing framework of the mobile army. For all the earlier declarations about the specialization of coast artillery service, the war demonstrated that coast artillery troops were generalists at heart.

The aging Weaver, whose memory and affections stretched over the entire span of modern coast defense, saw the participation of Coast Artillery Corps officers and men in every branch of the service as a grand endorsement of the virtues of coast defense training. If they had to, he offered, they could even go aboard the Navy's battleships and soon learn to operate the guns there.[23] His enthusiasm was nurtured by his absence from the European battlefields, and it was not shared by those who experienced the varied assignments of the coast artillery. Those officers who had been in France did not want to return to more fortification detail; they wanted to be shut of fixed defense entirely.

The critical flaw in the Coast Artillery Corps was the absence of any recent examples that justified its existence. Proponents of coast defense had cast about for events that proved its worth and sought them in recent conflicts—the Spanish-American War, the Russo-Japanese War, and the World War—but to no avail. Nothing demonstrated the conclusive value of coast artillery. There were no naval attacks of consequence on fixed defenses. The coast batteries in Belgium and in the Dardanelles seemed to point toward validation, a hope dashed by others who were quick to emphasize that those defenses indicated the value of mobile artillery rather than fixed guns in concrete batteries.[24] Career military men did not want to belong to a service that would never see action and that was doomed to waste away.[25] Their opinion of duty in the coast defenses was much like that expressed before the war by a British newspaper: "[t]here is no further role for the typical Garrison Gunner, rubicund of face, rotund of figure, wide of waist, seated on a stool in a position finding cell, watching the tide come in and go out, and waiting for ships which never pass by."[26]

After the war, members of the coast artillery proclaimed themselves to be part of the mobile army with a vehemence equaled only by their protests of twenty years earlier that they should be organized as a separate corps. Leading the insistent clamor was Major General F. W. Coe, the new chief of coast artillery. In what would have been considered heresy a few years before, Coe attacked the continued existence of the Coast Artillery Corps apart from the balance of the army. All arms have a place in coast defense, he said, and by the same token, coast artillery troops have a concern for the work of the army as a whole. Coe also felt that there was no fundamental difference in the basic artillery knowledge that every artillery officer should have since all guns shoot at targets. It did not matter if the weapon was bolted to a concrete emplacement or was cradled between rubber tires. For Coe, the major difference was not the

type of mount but the size of the gun; he proposed an organization based on light and heavy artillery, together within the same corps. His was a pan-service interpretation of coast defense in which the fortifications so painfully shepherded to completion were "nothing more than strong points for the line of infantry rifles and bayonets, which present the real defense against invasion."[27]

Behind the arguments was a passion for mobility and the need to escape the stigma acquired by fixed defenses during the war. Coe was an ardent advocate of railway artillery, and saw only two instances where permanent mounts might be considered necessary: when sixteen-inch guns were to be emplaced, because they were too large and heavy to be maneuvered over the railroads; and on islands without rail connections to the mainland.[28] With the army shrinking in the early 1920s, and the army's appropriations small and not likely to follow expensive directions, the reliance upon mobile artillery as an economical and war-proven defense seemed certain of support. If the Coast Artillery Corps was to survive, it could not afford to be associated with fixed batteries and permanently emplaced cannon. As put by one career coast artillery officer, the Corps had "arrived at a crucial point in its history. Many of its tried and true mechanisms must be scrapped, many others must be revolutionized; otherwise the Corps will soon find itself at the tail of the procession instead of at the head where it belongs. And there is no time to lose, either; something must be done, and done immediately."[29]

Suddenly embarrassed by its dated weapons, the practitioners of coast artillery linked the future of the Corps with the destruction of its past. The period before the war was labeled a time of experimentation rather than an era that represented the full flowering of coast defense.[30] The chief of coast artillery suggested in 1920 that coast artillery should be regarded as "in the process of development."[31] There was, the rationale went, nothing to be gained from examining what had gone before. Instead, the pre-war years were vilified and portrayed as an unproductive time when right-thinking artillerists were trapped between the "armament policy of short, low-velocity, obsolescent guns, the queer offspring of the marriage of the bug-bear of erosion to the haunting ghost of the disappearing carriage [...and] the emplacement policy of open, unconcealed, two level obsolescent batteries."[32]

Because of such profound changes in attitude, there were equally profound changes within the existing structure of coast defense. Among the most visible was the reorganization of all coast artillery companies into coast artillery regiments in 1924.[33] The regimental format, as had been emphasized in the struggle to establish an appropriate organization for the new coast defenses, was the creature of the mobile army. It still had little applicability in coast defense—it had far greater utility out of the defenses than in them—and its use indicated that the unique place of coast artillery service had come to an end.

As if to underscore the diminished role, all the coast defenses were relabeled harbor defenses in 1925, a careful redefinition that might have been done to imply that the coast artillery was willing to share its burden with other branches of the army.[34] It was a needless gesture since no one, from the time of the Endicott Board on, believed that the network of defended harbors was in fact a means of defending the whole coast. The term "coast defense" had always been used loosely rather than specifically, and the post-war change was more another break with the past than an action of greater weight.

In its materiel, the new coast artillery was interested only in mobile cannon that could be used for land targets and ships alike, as well as anti-aircraft artillery. By 1921, the instruction at the Fort Monroe Coast Artillery School was devoted to these two types of artillery almost to the exclusion of the more traditional coast defense weapons.[35]

The concentration upon mobile artillery brought few changes to Puget Sound. When the five-inch guns were removed from Battery Warner in 1926, the same water area was assigned to a pair of 155mm guns, which were held at Fort Lewis and never appeared at Fort Ward.[36] Firing positions and grades for railway artillery were begun at Cape George, seven miles west of Fort

Worden, in 1925 but they were never completed.[37] The only railway artillery to be used in the region was limited to several eight-inch guns sited on the outskirts of Port Angeles during World War II.[38] By that time, there was little chance that any heavy warship would attempt to enter Admiralty Inlet. A raid by fast torpedo boats was more likely, and large numbers of 37mm and 90mm guns were trailered in during the early 1940s to prepare for the new threat.

The enthusiasm of proponents aside, weapons on mobile carriages could not become the core of a national coast defense system. Unlike cannon mounted in concrete batteries, it took hours to prepare a railway gun for firing; some of the larger calibers could be used only on permanent firing foundations built during peace time at selected intervals along a length of track. Railway artillery offered little real mobility and sacrificed the protection, accuracy, and convenience afforded fixed coast cannon. Its eager promotion by members of the Coast Artillery Corps and Congress was inspired by the limited funds available after World War I, a period when all the military services struggled to maintain themselves. A mobile coast defense was popular in the 1920s and '30s because there was no money for anything else; when construction dollars became available following the outbreak of World War II, it was again fixed artillery that had primacy. Despite that brief resurgence, the Coast Artillery Corps had been irrevocably changed by its brush with mobility and by the late 1930s, it had become far more of an anti-aircraft service than a legion of heavy artillery specialists.[39]

Newer weapons in more modern emplacements, the result of war-time spending in the 1940s, surpassed the cannon that had protected Puget Sound since the turn of the century. Batteries 131 and 249 at Striped Peak, better known as Camp Hayden, mounted their guns in camouflaged emplacements with heavy overhead protection, an acknowledgement of the significance of the airplane in determining the design of contemporary seacoast fortifications.[40] The older batteries had become conspicuous targets from the air, casting bold shadows and prominent outlines which no disguise could mask completely.[41] There was some interest in modifying a few of the emplacements to protect them against aerial bombardment, but the initial investigation of Battery Benson, selected as a test case, was so disappointing that the project was given up. The necessary structure would have been complex and inefficient, not to mention expensive; the conclusion of the report was that more could be gained by mounting the guns in new batteries built somewhere else.[42] Valueless for their original purpose, the heavy guns were sold for scrap. The anti-motor torpedo boat guns, coupled with the remaining three-inch batteries, were the final remnant of the once formidable barrier at Admiralty Inlet. They were soon gone, and so too was the Coast Artillery Corps, abolished in 1950.[43]

Following the end of coast defense, the fortifications in Puget Sound began a transition. During the Korean War, the reservations at Admiralty Inlet became the site of training activities for an amphibious engineer unit; the army and navy continued the limited use of Fort Worden into the 1960s. For all of these purposes, the fortifications were of no consequence. Their prominence as irreplaceable elements of an important barrier was long past; they were no more than curious derelicts that disfigured the landscape. A dense thicket closed over the ignored fire control stations on the hill behind the Fort Casey main battery. There was no one to clean the drains in the heavy gun emplacements at Fort Flagler and they stopped up, rainwater flooding the magazines and storage rooms. Scotch broom, alder, and tangles of blackberry vines grew quickly and soon roads that were once busy with the traffic of horses and wagons were passable only by rabbits. Battery Lee disappeared behind the heavy green mask, and so did Batteries Schenck and Seymour, and Nash and Thornburgh, and almost every base end station and searchlight position.

The Washington State Parks and Recreation Commission acquired most of the acreage that constituted the old coast defenses. Park employees and volunteers cleared the brush and cleaned the trash from the abandoned emplacements. Choice recreation sites, the one-time military reservations attracted a new population, one that was not concerned with the cultural significance of such places or what they might retain as examples of engineering skill. No battle had been fought in Puget Sound, no invasion had been thwarted. Casual visitors, and many of those charged with the management of the parks themselves, believed that historical importance was something that could be measured by events of magnitude, things that people remembered and talked about. For all those soldiers that had served the guns for all those years, there was little that could be said. Former artillerymen would sometimes return with their families and offer a few recollections—the first kitchen police, an exciting target practice, an abrasive sergeant—to summarize the contribution made by the defenses. With that interpretation, the surviving parts seemed the foolish legacy of an age whose decisions must have been made in a cloud of hopeless innocence, unremarkable artifacts without value, tolerated because of their ponderous bulk.

Also with that interpretation, we are left with no guidance at all to help us understand what we are to make of these remnants. Are we willing to believe and accept that what many minds and hands built long ago is without meaning? The answer to that question must be no and that answer alone encourages us to look for proof of value. We can find guidance in three places: in the opinions of the men who served during the early years of the defenses when they were most important, in the perspective of the builders themselves, and in the view of scholars who know something about the history of fortification. What we learn may surprise us.

Plate 7-1. Fort Casey Battery Trevor in 1923, with the empty emplacements of Battery Valleau to the right. There were six examples of this same design in the Admiralty Inlet fortifications. The guns of Battery Trevor were removed and sent to Corregidor ten years after this photo was taken; the other like batteries retained their weapons until after World War II and constituted then all that was left of the former Coast Defenses of Puget Sound. In the end their small three-inch guns were more useful than the big cannon that had been the center of attention at the beginning of the 20th century, a shift of emphasis brought about by a change in the potential threat posed by small fast torpedo boats rather than the big armored warships of the 1890s. *Steven Kobylk collection*

Epilogue

The speech given in 1926 by Percy Kessler, the commander of the Puget Sound defenses, opened chapter one of this work. It is now time to return to Colonel Kessler to see what he might have to tell us about the lasting significance of the fortifications; we need to bring his story, and our story, to an end.

Kessler was an accomplished officer of the Coast Artillery Corps. Before he was posted to Puget Sound, he had been to the submarine mine school at Fort Totten, New York; served as a Fire Commander at Fort Monroe, Virginia; named as the commanding officer of Fort Strong in Massachusetts; and ordered to San Diego to take charge of the defenses protecting that city. He graduated from the Army War College, served two tours in the fortifications guarding the Panama Canal, and was variously the commanding officer of Fort Winfield Scott in San Francisco and Fort Mills on Corregidor Island in the harbor of Manila. His wish was to end his career in command of Fort Hancock, part of the defenses of New York, and there he was sent in 1934. He died suddenly the next year after returning home from a movie at the post theater.[1]

Given those many assignments, Kessler was well placed to make observations about the fortifications, armament, techniques, and people who made and manned the American system of coast defense. Before his appointment to the military academy at West Point, he had begun engineering studies at Purdue, so he may also have had more than passing interest in the design and construction of the structures of the defense. Yet if he had an opinion about it all—and it is hard to believe that he did not—he made no record of it. He may have thought about doing so since he bought a journal; it survives to this day, mostly blank. Even the entry for such a momentous occasion as the birth of his son reads only "baby boy."[2] As it was with almost every individual named in this book, Kessler made no lasting account of his own take on things, no comment on what was happening around him, no insights that might help others understand and assess the importance of it all and how it might fit into the larger scale of military architecture, the history of fortification, and the history of the army.

Given the silence of so many whom we most wish would speak, it is unexpected and welcome that one voice was raised on the question. Eben Eveleth Winslow, an officer of the Corps of Engineers, lectured on the design and construction of seacoast fortifications. He believed that fortifications had historical qualities, and in 1917 he wrote to the army engineers in various harbors and encouraged them to rehabilitate and care for important examples so that they could be maintained for their significance. It does not appear that anyone was moved by his proposal.[3]

Winslow did not explain what he thought an important example of a coastal fortification might be and he did not detail what aspects of them might shape the question of significance. It is likely that the idea of historical significance, as many use the term today, would have been alien to him. In the last fifty years in the United States we have come to understand that significance can be found in a dazzling array of objects, sites, and buildings, from the conventionally splendid mansions of great men to the log cabins of the obscure, from wooden-hulled sailing ships to steam-era railroad stations and the quirky shapes of atomic age motels and auto courts. In all probability, Winslow was not thinking of fortifications as important in a broad cultural way, but more along the lines of what certain designs and structures could tell his peers about improving designs yet to be made and structures yet to be built. An historically important fortification to Winslow was apt to be one whose study could result in better fortifications. As he himself said, the development of fortification was "a sort of evolution, based upon experience that has shown which of the many things attempted in the past should be retained and which should be discarded." For him, ultimate value was banked in utility:

> [i]f our seacoast fortifications have been so well designed as to have deterred an enemy from attacking them, to have kept him out of a harbor plentifully supplied with landing and dock facilities, and to have forced him to land elsewhere under difficult conditions, and where proper terminal facilities are lacking, they must be considered to have performed most efficiently the main purpose for which they were designed. Nothing more should be desired or expected of them. In fact, a seacoast fortification may be said to have most efficiently performed the function for which it was intended, if it is never called into action at all.[4]

His argument is not unique to fortifications, however, and it doesn't help us much in our search for significance. The conclusion applies to most military expenditures; the investment is made as a form of insurance, and the assumption is that once the investment has been made, it has of itself a deterrent effect. We cannot easily test or measure that point of view, but it seems to make sense.

For the most part then, engineers like Winslow, who designed and built the defenses, and artillery men like Kessler, who conjured an effective defense out of an eclectic imagination, have not much to tell us about what their work might represent in our present. The task is made more difficult by the reluctance of many—most—scholars to assign much importance at all to the products of the Endicott Board. It is their preference to embrace earlier coastal fortifications as capturing all that is worthwhile in the heritage of domestic fortifications. Some authors talk about a golden age of American fortification, and almost without exception they mean the years and products of the period before the Civil War. One writer has called it the "last great period of innovative fortification…[producing] the most elaborate and, historically, most interesting forts of American history," and another author referred to it as a "stable refuge in a fast-changing world," suggesting further that it dominated the thinking of military engineers and left them helpless in the face of an increasingly industrial age that required other forms of fixed defense. A writer in an excellent treatment of masonry fortifications concluded his presentation by noting that after the Civil War, "architecture for defense was simply unable to keep pace with weapon design," observing further that in considering the aesthetic qualities of masonry works, their "beauty of form result[ed] from clarity of purpose," implying that later defenses lacked the same clarity. These observations miss the revolutionary creativity that is at the core of the movement to construct new defenses as the nineteenth century drew to a close.[5]

The remarks make clear that for some people the form of the earlier fortifications constitutes the major element of significance and that they have a preference for the aesthetic represented by the impressive volumes and details that characterize coastal forts built of brick and stone masonry. We can all agree that the changes in fortification design brought about by the adoption of high-powered ordnance in concrete batteries were dramatic. Certainly the members of the Corps of Engineers saw the changes as profound and could scarcely believe the extent of the transformation. "We must sacrifice neat crests and beautiful slopes," said H. L. Abbott, one of the most influential figures in the modernization of the nation's defenses. "Trees and bushes must be planted on the parapets and behind the batteries to prevent a clear definition of the guns; the batteries themselves must be colored to harmonize with their surroundings in summer and winter."[6] By the end of the 1860s, the three basic qualities of all masonry fortifications—familiarity of materials, familiarity of presence, and familiarity of scale—disappeared and the fortress silhouetted against the evening sky remained only as a romantic image with no military value. And however seductive it might be, romantic imagery is not a useful indicator of broader significance.

The fundamental purpose of those brick and stone fortresses had not changed with the coming of the creations spurred by the Endicott Board; the intent was still to defend important ports. The means had not changed; the basic weapon was still the heavy cannon, albeit one that was vastly improved. The shift was in the dispersal of different elements of the fortification that transformed the harbor entrance into an extensive military landscape. The fortress of old had become atomized and was now an aggregate of specialized components scattered throughout large parcels of real estate, all connected by telephone. It was a whole of many parts, all born of a flood of technology that coursed through the nation at the close of the Civil War.

A coastal fort had become a web of emplacements, base-end stations, searchlights, and other structures that in their form tended to be architectonic rather than architectural. Details were few, and in the largest of them the strongest impression was one of a repeating pattern of simple geometric shapes, sometimes relieved by recesses or doors and shutters of steel that contrasted to great effect on the otherwise featureless planes. Craftsmanship survived. It is easy to recognize the skill of a stone mason, and those who worked in concrete were akin to them in that they sought results that were handsome. For example, concrete finishers took great pains to smooth over the coarse outlines left by the form boards, producing a surface that was clean and unblemished. Designers also made deliberate choices based on a desire for the fortifications to look "right" and to satisfy the emotional involvement in building that is one of the qualities of architecture. These were modest decisions—tapering a column, changing the proportions of a room, rendering an element to suggest greater strength—but they were impactful and particularly apparent in Puget Sound.

Not everything that was built in the 1890s and early 1900s exhibited the talents of a skilled designer or tradesman, but much of the fortification construction in Puget Sound during the period did exactly that. It recalls the origin of the term "engineer" in the Latin word "ingenium," meaning a clever thought or invention, and its progression over the centuries to combine a sense of art, craft, and science.[7] It helps to keep in mind the observation of writer and engineer Samuel C. Florman that "every manmade structure, no matter how mundane, has a little bit of the cathedral in it, since man cannot but transcend himself as soon as he begins to design and construct."[8] There is a singleness of purpose in cathedral and fortification alike, as we have seen in these pages, and the creative pulse that can produce a cathedral can also bring forth an elaborate system of fortification that is both structure and landscape. We are more at ease in finding reasons to praise the architecture of cathedrals than we are the architecture of fortifications that are little more than one hundred years old, yet the practice and the reasons to do so are the same since they spring from a common sense of creating carefully from a grand design.

Of course the cathedral analogy is just a device to help us understand the complexity and heritage implicit in structures of many kinds and it has limitations, one of the most obvious being duration: the fortifications in Puget Sound were in use for a very short period indeed when compared to great houses of worship, underscoring the fact that durability of materials does not necessarily imply durability of function. One of the compelling observations about the Endicott fortifications is how quickly their years of effective use seemed to pass, and perhaps that is what they have to teach us, that what is competent one moment may be out of date in the next. The battleship as it emerged in the 1870s was the threat in mind when the fortifications of the 1890s began to take shape, and although the fixed defenses could initially prevent such vessels from forcing their way through Admiralty Inlet and similar waters elsewhere, it was an advantage that slipped away in the face of continuing change in naval battle fleets. There was another factor, too. In the early years of the twentieth century, domestic policy moved in a way that supported a more powerful navy, one strong enough to reduce the dependence on the army's collection of guns,

mortars, and mines. Shifts in technology and shifts in circumstances worked against an enduring elaboration of coastal defense.[9]

At his death, Kessler's body was placed aboard a coast defense
mine planter and taken up the Hudson River for burial at West
Point. Every graduate of the military academy can receive a
gold class ring, a ring that is usually worn for a lifetime. There is
a tradition that at the end of that lifetime, the ring is returned
and melted down, and the pool of metal becomes the source of
gold for new rings that are cast for new graduates. The practice
symbolizes continuity and transition, and it began with Percy
Kessler; it was his ring that was the first to be added to the molten
reservoir. The legacy of the Puget Sound fortifications also speaks
to continuity and transition. They represent the most complete
and sophisticated system ever to defend the important harbors
of the United States, and in many ways they are the culmination
of a tradition of design and construction that reaches back to the
walls of Jericho. For the men like Kessler and others who manned
the defenses, they represent something else as well: the transition of the army from the post-Civil War doldrums of spiritless
police duty to a new century and a new and important role in the
defense of an expanding nation.

Appendix
Armament Summary

The following tables display basic information about the construction and disestablishment of the gun and mortar batteries in the defenses. Most of the dates are those included in the correspondence or reports prepared at the time rather than the Army's "Report of Completed Works" documents that were developed much later. Some of the sources did not include a date of the month, and to be consistent throughout the table, only the month and year is given.

In the "Battery and Armament" column, the date refers to the model year of the carriage mounting the cannon. Model year designations were also given to cannon to indicate changes in their characteristics, but those changes were often less noticeable than those associated with the mounts.

LEGEND

DC — Disappearing Carriage
BC — Barbette Carriage
AGL — Altered Gun Lift
Ped — Pedestal
BP — Balanced Pillar
Mr — Mortar

Fort Casey

Battery and armament	Construction		Transferred to Artillery	Named	Armament		Notes
	Begun	Complete			Installed	Removed	
William Worth 2 – 10" DC 1896	Aug. 1897	Dec. 1898	June 1902	Jan. 1906 *b*	1900	1942	*a* First name not included in official designation.
James Moore 3 – 10" DC 1896	Aug. 1897	1904	June 1902 *c*	Dec. 1904	2 – 1902 1 – 1909	1942	*b* Part of Battery Moore until that date. *c* Emplacement no. 3 transferred November 1905.
Henry Kingsbury 1 – 10" DC 1896 1 – 10" DC 1901	Mar. 1901	Nov. 1905 *d*	Nov. 1905	Dec. 1904	1 – 1903 1 – 1906	*e*	*d* Although reported completed at the date shown, modifications to emplacement no. 2 continued until September 1906.
Truman Seymour *a* 8 – 12" Mr 1896MI	Apr. 1898	Mar. 1899	June 1902	June 1903	Dec. 1900 *f*	*g*	*e* Both guns of the battery were apparently removed in June 1918. In 1919 a spare gun already at Fort Casey was remounted in emplacement no. 2. The carriage in emplacement no. 1 was scrapped November 1920, and no. 2 gun and carriage were scrapped in 1942.
Alexander Schenck 8 – 12" Mr 1896MI	Apr. 1898	Mar. 1899	June 1902	Jan. 1906 *h*	Dec. 1900	1942	
John Valleau 4 – 6" DC 1903	Oct. 1903	Mar. 1907	May 1907	Dec. 1904	Aug. 1908	Nov. 1917 *i*	*f* Four cannon mounted by artillery troops somewhat later.
Thomas Parker 2 – 6" DC 1903	Aug. 1903	Aug. 1905	May 1907	Dec. 1904	Aug. 1908	Nov. 1917 *i*	*g* Two forward cannon removed from each pit May 1918, for railway mounting; carriages scrapped December 1920. Balance of armament scrapped 1942.
Reuben Turman 2 – 5" BP 1896	Aug. 1899	June 1901	June 1902	Dec. 1904	1902	May 1918 *j*	*h* Part of Battery Seymour until that date.
John Trevor 2 – 3" Ped 1903	Aug. 1903	June 1905	May 1907	Dec. 1904	July 1908	Nov. 1933 *k*	*i* Removed for wheeled mounts; carriages scrapped November 1920.
Isaac Van Horne 2 – 3" Ped 1903	Aug. 1903	June 1905	May 1907	Dec. 1904	July 1908	*l*	*j* Sent to Sandy Hook Proving Ground, New Jersey; carriages scrapped 1920. *k* Guns shipped to Corregidor and carriages scrapped 1942. *l* Salvaged sometime after March 1946.

FORT FLAGLER

Battery and armament	Construction		Transferred to Artillery	Named	Armament		Notes
	Begun	Complete			Installed	Removed	
William Wilhelm *a* 2 – 12" AGL	Aug. 1897	Mar. 1899	Aug. 1902	Feb. 1902	Nov. 1899	*b*	*a* First name not included in official designation.
John Rawlins 2 – 10" BC	Aug. 1897	Mar. 1899	Aug. 1902	Jan. 1906 *c*	Nov. 1899	June 1918 *d*	*b* Salvaged pursuant to order of October 1942.
Paul Revere 2 – 10" BC	Aug. 1897	Mar. 1899	Aug. 1902	Jan. 1906 *c*	Nov. 1899	Apr. 1941 *e*	*c* Separated from Battery Wilhelm at that date.
Henry Bankhead 8 – 12" Mr 1896MI	Jan. 1901	June 1902	Aug. 1902	Dec. 1904	1902	*f*	*d* Guns taken away for railway mounting; carriages scrapped October 1920. *e* Armament transferred to Canada.
John Grattan 2 – 6" DC 1903	Sept. 1904	June 1906	Apr. 1907	Dec. 1904	1 – Apr. '07 1 – May '07	Nov. 1917 *g*	*f* Two forward cannon removed from each pit, May 1918; carriages scrapped October 1920. Balance salvaged December 1942.
James Calwell 4 – 6" DC 1903	June 1903	July 1905	Apr. 1907	Dec. 1904	3 – Mar. '06 1 – May '07	Nov. 1917 *g*	*g* Guns taken for wheeled mounts; carriages scrapped October 1920.
Walter Lee *a* 2 – 5" BP	July 1899	Nov. 1900	Aug. 1902	Feb. 1902	1901	Oct. 1918 *h*	*h* Guns transferred to Westport and remounted. *i* Salvaged subsequent to order of March 1946.
Edward Downes 2 – 3" Ped	Aug. 1903	Sept. 1905	Apr. 1907	Dec. 1904	1 – Oct. '09 1 – Jul. '10	*i*	
Thomas Wansboro 2 – 3" Ped	Aug. 1903	Sept. 1905	Apr. 1907	Dec. 1904	July 1908	*i*	

Fort Ward

Battery and armament	Construction		Transferred to Artillery	Named	Armament		Notes
	Begun	Complete			Installed	Removed	
Francis Nash 3 – 8" DC 1896	Mar. 1900	1901	Jan. 1904	Dec. 1904	Apr. 1903	Oct. 1917 *a*	*a* Guns taken for railway mounting; carriages scrapped January 1920.
William Warner 2 – 5" Ped 1903	1900	Jan. 1904	Jan. 1904	Dec. 1904	Oct. 1907 Mar. 1919	Oct. 1917 *b* June 1926	*b* Originally removed for use on the transport ship *Dix*, but installation was never made. Remounted for dates shown.
Thomas Thornburgh 4 – 3" BP 1898	1900	Oct. 1903	Jan. 1904	Dec. 1904	Aug. 1904	July 1920 *c*	*c* Armament declared obsolete March 1920; carriages scrapped December 1920.
John Vinton 2 – 3" BP 1898	1900	Oct. 1903	Jan. 1904	Dec. 1904	Aug. 1904	July 1920 *c*	*d* Armament never installed.
William Mitchell 2 – 3" Ped 1903	1900	Oct. 1903	Jan. 1904	Dec. 1904	*d*		

Fort Whitman

Battery and armament	Construction		Transferred to Artillery	Named	Armament		Notes
	Begun	Complete			Installed	Removed	
G. F. E. Harrison *a* 4 – 6" DC 1905 MI	1909	1911	May 1911	Dec. 1909	1910	*b*	*a* First name not included in official designation. *b* Salvaged pursuant to order of September 1943.

FORT WORDEN

Battery and armament	Construction		Transferred to Artillery	Named	Armament		Notes
	Begun	Complete			Installed	Removed	
Joseph Ash 2 – 12" *b* BC *c* 1896	Aug. 1898	June 1900	June 1902	Dec. 1904 1 – Oct. '00	1 – Apr. '02 *d*	Dec. 1942	
August Quarles 3 – 10"*e* BC 1893	Aug. 1898	June 1900	June 1902	Dec. 1904	1 – Feb. '01 1 – Jan. '00 1 – Dec. '00	Apr. 1941 *f*	
Alanson Randol 2 – 10"*g* BC 1893	Aug. 1898	June 1900	June 1902	Dec. 1904	Feb. 1901	June 1918 *h*	
John Brannan 8 – 12" Mr 1896MI	July 1899	Jan. 1901	June 1902	Dec. 1904	Aug. 1900	*i*	
James Powell *j* 8 – 12" Mr 1896MI	July 1899	Jan. 1901	June 1902	Jan. 1906	Aug. 1900	*k*	
Henry Benson 2 – 10" DC 1901	1904	July 1906	Apr. 1908	Dec. 1904	Feb. 1908	*l*	
David Kinzie *a* 2 – 12" DC 1901	Sept. 1908	1910	Jan. 1912	Dec. 1909	1912	*m*	
Cornelius Tolles 4 – 6" DC 1903	June 1903	Nov. 1906	May 1907	Dec. 1904	1906	*n*	
Tolles 'B' 2 – 6" BC 1900	Nov. 1936	June 1937	Aug. 1937	n/a	June 1937	1948	
Amos Stoddard 4 – 6" DC 1903	July 1903	Nov. 1906	May 1907	Dec. 1904	Oct. 1907	Nov. 1917 *o*	
Thomas Vicars 2– 5" BP	Apr. 1900	Sept. 1900	June 1902	Dec. 1904	1902	Nov. 1917 *o*	
Haldimand Putnam 2 – 3" Ped	June 1903	Nov. 1906	May 1907	Dec. 1904	June 1908	*p*	
Samuel Walker 2 – 3" Ped	June 1903	Nov. 1906	May 1907	Dec. 1904	1 – Feb. '10 1 – June '10	*p*	

a First name not included in official designation.

b Originally one 10"; in 1905, armament exchanged with 12" of Battery Quarles and in 1906, the battery incorporated the 12" gun previously assigned to Battery Powell (see note *j*).

c Originally included one AGL; replaced with BC February 1909.

d 10" gun and carriage removed May, 1905, from emplacement no. 2 and replaced with 12" gun and carriage from Battery Quarles.

e Originally one 12"; became 10" after 1905 exchange with Battery Ash. Emplacements added from Battery Randol in January 1906, and March 1909.

f Transferred to Canada.

g Originally four 10"; emplacements 3 and 4 transferred to Battery Quarles.

h Guns taken for railway mounting. One carriage scrapped August 1919, one retained for parts.

i Two forward mortars removed from each pit for railway mounting May 1918; carriages scrapped December 1920. Remaining armament salvaged pursuant to order of February 1944.

j Name originally assigned to main gun battery; transferred to pits C and D of Battery Brannan in 1906.

k Two forward mortars removed from each pit for railway mounting, May 1918; carriages scrapped December 1920. Remaining armament salvaged December 1942.

l Salvaged pursuant to order of June 1943.

m Salvaged pursuant to order of April 1944.

n Guns 3 and 4 removed November 1917, for wheeled mounts; carriages salvaged December 1920, and emplacements later designated as Tolles 'B'. Guns 1 and 2 (Tolles 'A') removed June 1943; carriages salvaged January 1944.

o Guns taken for wheeled mounts; carriages salvaged December 1920.

p Salvaged pursuant to order of March 1946.

Selected Bibliography

There are few secondary works that treat the history, technology, or architecture of American coast defenses, and descriptions of the Coast Defenses of Puget Sound are even more limited. For the most part then, the author has depended upon original sources, including a good many that are rare or not easily available. The most pertinent have been cited in the footnotes. The bibliography presents the sources that were the most useful, and in many instances, the most likely to be in the collection of a major library.

Chief of Artillery, United States Army. Annual Reports, 1901–1908.

Chief of Coast Artillery, United States Army. Annual Reports, 1909–1923.

Chief of Engineers, United States Army. Annual Reports, 1890–1920.

Chief of Ordnance, United States Army. Annual Reports, 1885–1908.

Hines, Frank T. and Ward, Franklin W. *The Service of Coast Artillery*. New York: Goodenough and Woglom Company, 1910.

Journal of the United States Artillery, Volumes 1–55, January 1892–October 1921.

Lewis, Emanuel Raymond. *Seacoast Fortifications of the United States*. Washington: Smithsonian Institution Press, 1970.

National Archives and Records Administration. Record Group 77. "Records of the U. S. Army Corps of Engineers, Seattle District." Boxes 241–332 (1890–1945).

_____. Record Group 392. Records of U. S. Army Coast Artillery Districts and Defenses, 1901–1942. Entry 286. "History of the Northwest Sector."

U. S. Congress. House. *Report of the Board of Fortification or Other Defenses*. House Executive Document 49, 49th Congress, 1st Session, 1886.

_____. Senate. *Coast Defenses of the United States and the Insular Possessions*. Senate Document 248, 59th Congress, 1st Session, 1906.

Winslow, Eben Eveleth. *Notes on Seacoast Fortification Construction*. Number 61. Occasional Papers of the Engineer School, United States Army. Washington: Government Printing Office, 1920.

Notes

The following abbreviations have been used throughout the citations:

ARCE: Annual Report of the Chief of Engineers
ARCO: Annual Report of the Chief of Ordnance
NARA: National Archives and Records Administration
JUSA: Journal of the United States Artillery
ARSW: Annual Report of the Secretary of War
ARCA: Annual Report of the Chief of Artillery
ARCCA: Annual Report of the Chief of Coast Artillery

Chapter 1

1. "Forts Undermanned, Says Colonel Kessler," *Seattle Times*, April 3, 1926, 3.
2. "Puget Sound Defense Force Smallest in 25 Years," *Seattle Post-Intelligencer*, April 3, 1926, 9.
3. National Archives and Records Administration [hereafter cited as NARA], Record Group 77, "Records of the Office of the Chief of Engineers Pertaining to Puget Sound," Box 300, File 345.1, Folder No. 4, John C. Phillips to Percy Kessler, April 1, 1926 [materials from this source hereafter cited only by box number].
4. Ibid.
5. Senate Executive Document 165, 50th Congress, 1st Session, (Serial 2513), 63 [hereafter cited as the Puget Sound Reports].
6. Ibid., 54.
7. Ibid., 58.
8. Ibid., 55.
9. Ibid., 61.
10. Ibid., 41.
11. Ibid., 68.
12. Ibid., 32.
13. Ibid., 30-40.
14. Puget Sound Reports, 36 and 52.
15. Ibid., 52.
16. "Harbor of Refuge at Neah Bay," *Port Townsend Leader*, October 5, 1899.
17. Emanuel Raymond Lewis, *Seacoast Fortifications of the United States* (Washington: Smithsonian Institution Press 1970), 76.
18. *Congressional Record*, 52nd Congress, 1st Session (Volume 23, Part 7), July 5, 1892, 6204.
19. John D. Allen, *The American Steel Navy* (Annapolis: Naval Institute Press, 1972), 13.
20. Lewis, *Seacoast Fortifications*, 77.
21. House Executive Document 49, 49th Congress, 1st Session (Serials 2395 and 2396), report dated January 23, 1886 [hereafter cited as the Endicott Board Report].
22. Ibid., 6.
23. Ibid., 28.
24. See data presented in the Annual Statement of the Chief of the Bureau of Statistics on the Commerce and Navigation of the United States (Serials 1966, 2025, 2107, and 2197).
25. Endicott Board Report, 253.
26. NARA, Record Group 392, Item 286, "History of the Pacific Coast Defenses," 77 [also known by its subsequent title of "History of the Northwest Sector"]; Endicott Board Report, 188.
27. Puget Sound Reports, 18.
28. Ibid., 19.
29. Russel F. Weigley, *History of the United States Army* (New York: MacMillan and Company 1967), 289. Miles declared his suspicion of the disappearing carriage in a form readily consumable by the press and, while General of the Army, insisted upon the installation of several newsworthy pneumatic dynamite guns over the objections of both the Chief of Engineers and the Chief of Ordnance.

30. Nelson A. Miles, *Personal Recollections of General Nelson A. Miles* (Chicago and New York: Werner Company 1896), 402.

31. Puget Sound Reports, 16.

32. Ibid.

33. *Congressional Record*, 49th Congress, 1st Session (Volume 17, Part 3), March 1, 1886.

34. Puget Sound Reports, 2.

35. Senate Report 1213, 51st Congress, 1st Session (Serial 2709), 105 and 110.

36. *Congressional Record*, 52nd Congress, 1st Session (Volume 23, Part 7), July 15, 1892, 6204.

37. Senate Report 321, 52nd Congress, 1st Session (Serial 2519), 2.

38. Edmund S. Meany, *Governors of Washington: Territorial and State* (Seattle: University of Washington 1915), 60-61.

39. *Congressional Record*, 51st Congress, 2nd Session (Volume 22, Part 2), February 3, 1891, 2089.

40. Ibid., 2087.

41. *Congressional Record*, 53rd Congress, 2nd Session (Volume 26, Part 3), March 2, 1984, 2535.

42. House Executive Document 1, 51st Congress, 1st Session (Serial 2721), Annual Report of the Secretary of the Navy, 1889, "Report of the Commission to Select a Site for a Navy-yard on the Pacific Coast North of the Forty-second Parallel of North Latitude," 139.

43. Ambrose Barkley Wyckoff, "Starting the Puget Sound Navy Yard and the Dry-dock, and The actual beginning of the Lake Washington canal," typescript, November 6, 1908, 4.

44. Annual Report of the Secretary of the Navy, 1889, "Report of the Commission…," 143.

45. Senate Executive Document 24, 51st Congress, 2nd Session, (Serial 2818), "Message from the President of the United States transmitting a report of the Commission to select a site for a dry dock on the Pacific coast," 14.

46. History of the Pacific Coast Defenses, 82.

47. Ibid.

48. Ibid., 83.

49. Ibid., 86.

50. Allen, *American Steel Navy*, 4.

51. *Congressional Record*, 53rd Congress, 3rd Session (Volume 27, Part 4), February 28, 1895, 2897.

Chapter 2

1. *Leader*, March 5, 1896.

2. James G. Swan, "Puget Sound's Fortifications," *Leader*, June 18, 1896.

3. Swan, "Interesting Talk with General Miles," *Leader*, October 1, 1896.

4. Swan, "Early History of Fortifications," *Leader*, December 24, 1896.

5. Ibid.

6. Swan, *Leader*, June 18, 1896.

7. Swan, *Leader*, October 1, 1896.

8. Box 385, File 3021, undated note of Harry Taylor.

9. C. H. Hanford, *Halcyon Days in Port Townsend* (Seattle: privately printed, 1925), 90.

10. *Island County Times*, April 1, 1898.

11. House Document No 2, 56th Congress, 1st Session, *Annual Report of the Chief of Engineers*, 1899, Serial 3905, 4 [hereafter cited as *ARCE*].

12. Robert Stanton Browning III, "Shielding the Republic: American Coastal Defense Policy in the Nineteenth Century," University of Wisconsin-Madison, Ph.D. dissertation, 1981, 289.

13. Allen Johnson, ed., *Dictionary of American Biography* (New York: Charles Scribner's Sons, 1928), 327-28.

14. Gordon B. Dodds, *Hiram Martin Chittenden* (University of Kentucky, 1973), passim.

15. E. E. Winslow, *Notes on Seacoast Fortification Construction* (Washington: GPO, 1920), 348.

16. Personal Notes, May 29, 1908, and diary entry for May 3, 1908, Chittenden Papers, Washington State Historical Society, Tacoma, Washington; see also *Dictionary of American Biography*, 77-78.

17. Herbert Hunt, *Tacoma: Its History and Its Buildings* (Chicago: S. J. Clarke Publishing Co., 1916), 505-508.

18. *Leader*, March 16, 1923.

19. Caroline Ober Papers, University of Washington Library Special Collections.

20. Author interview with H. E. McMorris, May 10, 1976.

21. Box 285, File 2905, R. H. Ober to John Millis, July 28, 1904.

22. Box 249, File 347-798, Ricksecker to Taylor, July 2, 1896.

23. Box 249, File 347-798, Ricksecker to [no first name] Geary, June 8, 1896.

24. Box 249, File 347-798, Ricksecker to Taylor, July 2, 1898.

25. *ARCE* 1897, Serial 3631, 9.

26. Ibid.

27. Box 249, File 805-1279, Taylor to Chief of Engineers, October 28, 1896.

28. *ARCE*, 1897, Serial 3631, 764.

29. John J. Maney, Albert C. Goerig, and Arvid Rydstrom. Maney and Goerig were Everett businessmen. Maney advertised himself as a bridge, wharf, and building contractor. He erected the Everett nailworks, that community's first major industry of the 1890 boom period. Arvid Rydstrom was a former commissioner of public works in Tacoma. The three men apparently formed the partnership especially for the defense work, and dissolved it shortly after they completed their contracts.

30. Box 291, File 1-52, Ricksecker to M. L. Walker, August 10, 1897.

31. Box 291, File 86-117, Progress Report for the Month of August, 1897.

32. Box 250, File 61-126, Progress Report for the Month of February, 1898.

33. Box 250, File 61-126, Progress Report for the Month of January, 1898.

34. Box 250, File 61-126; the conflict is outlined well in a series of letters from the major protagonists in April 1898.

35. Box 252, File 135-180, Ricksecker to Walker, February 1, 1898.

36. Box 293, File 461-500, Philip Eastwick to Maney, Goerig and Rydstrom, March 26, 1898; Report of Operations for the Week Ending March 19, 1898.

37. Box 295, File 989-1050, Eastwick to Walker, November 12, 1898.

38. *ARCE*, 1899, Serial 3905, 1005 and 1008.

39. Box 251, File 1-53, Eastwick to Walker, January 5, 1899.

40. Box 250, File 262-346, Swigart to Taylor, June 9, 1898.

41. Box 250, File 262-346, Swigart to Taylor, May 7, 1898.

42. Box 252, File 54-134, Swigart to Taylor, January 17, 1899.

43. Ibid.

44. Box 253, File 310-365, an undated report examining the inventory after purchase.

45. Box 253, File 365-435, Maney to Taylor, March 28, 1899.

46. *Island County Times*, September 26, 1898.

47. "For The Defenses," *Leader*, February 18, 1897.

48. Box 252, File 54-134, Taylor to John Wilson, December 19, 1898.

49. *ARCE,* 1899, Serial 3905, 1009.

50. Box 252, File 181-215, Taylor to Wilson, February 16, 1899.

51. Box 249, File 1009-1299, Ricksecker to Taylor, August 5, 1897.

52. Box 249, File 1603-2016, Progress Report for the Month of October, 1897; Box 250, File 61-126, Progress Report for the Month of February, 1898.

53. Box 254, File 911-1000, D. W. McMorris to Walker, August 11, 1899; File 1001-1093, McMorris to Walker, August 22, 1899.

54. Box 254, File 1001-1093, McMorris to Walker, August 22, 1899.

55. Box 294, File 664-705, Report of Operations for the Month of June, 1898.

56. Box 291, File 1-52, Ricksecker to Walker, July 23, 1897.

57. Ibid.

58. Box 250, File 347-400, W. T. Preston to Walker, August 24, 1898.

59. Ibid.

60. Box 291, File 86-117, Ricksecker to Walker, September 6, 1897.

61. Box 253, File 366-435, undated inventory.

62. Box 285, File 2905, Ober to Millis, July 28, 1904; and Box 252, File 135-180, letter from Walker, date unrecorded.

63. Box 252, File 216-270, Progress Report for the Week Ending February 12, 1899.

64. Box 267, File 2568, Preston to Walker, April 20, 1900. Occasionally, supervisors and trustworthy employees were allowed to bring their families to the construction sites for the summer months; small shacks were built by the men to house their wives and children.

65. Box 250, File 262-346, Swigart to Taylor, June 9, 1898; and File 347-400, Swigart to Taylor, August 10, 1898.

66. Box 253, File 576-680, Preston to Walker, May 23, 1899.

67. Ibid.

68. Box 249, File 1301-1600, Progress Report for the Week of October 6, 1897; Box 252, File 181-215, Taylor to Wilson, February 16, 1899.

69. Box 249, File 1603-2016, Progress Report for the Week Ending November 20, 1897.

70. Box 252, File 181-215, Taylor to Wilson, February 16, 1899.

71. Box 272, File 2989, S.D. Mason and Ober to Millis, December 9, 1904.

72. Box 254, File 1001-1093, Preston to Walker, August 16, 1899.

73. Box 283, File 2503, report dated March 28, 1904.

74. Box 255, File 1412-1500, Preston to Walker, November 18, 1899.

75. Box 253, File 681-750, Preston to Walker, June 15, 1899. There were also occasional strikes a few years later when work was under-way to improve the post with roads and permanent buildings for the garrison.

76. Box 249, File 1603-2016, Progress Report for the Week Ending November 20, 1897.

77. Box 255, File 1327-1411, Thomas Dennehy to Elihu Root, October 6, 1899.

78. Box 253, File 751-830, Preston to Walker, Annual Report for the Fiscal Year Ending June 30, 1899.

79. Box 256, File 362-450, John Karcher to Taylor, March 30, 1900.

80. Ibid.

81. Box 253, File 437-500, Ricksecker to Walker, April 10, 1899.

82. Box 295, File 866-900, Preston to Walker, September 26, 1898.

83. Box 287, File 3195, Ober to Millis, August 19, 1904.

84. Ibid.

85. Record Group 92, Office of the Quartermaster General, Fort Worden, File 255530, NARA. Tramways that featured carts pushed by hand connected the Ordnance Machine Shop and several engi-neer buildings near the dock at Fort Worden; they were removed at an undetermined date. A similar tramway at Fort Casey appar-ently connected the post dock with the Quartermaster Storehouse, although little is known about its creation and removal.

86. Drawing F43-4-l, Fort Whitman Troop Housing, General Site Plan and Utilities, U.S. Engineer Office, Seattle, 2 June 1945, revised 2 August 1945, in author's collection.

87. Box 305, File 398.1, Report of the Artillery District Commander, July 20, 1907.

88. Box 252, an undated report examining the inventory.

89. Box 253, File 681-750, Mason to Walker, June 10, 1899.

90. "Office and Quarters at Point Wilson," April 1, 1899. Photo No. 18, Engineer Construction Photo Set, in author's collection.

91. Box 71, File 3148, Kutz to Preston, March 17, 1910.

92. "Buried Steam Locomotive Discovered at Fort Worden," *Leader,* February 22, 1973.

93. "Big Guns for Townsend's Fortifications," *Leader,* March 9, 1899.

94. "To Finish Work at Marrowstone," *Leader,* March 23, 1899.

95. Box 291, File 126-180, Completion Report for the Month of September, 1897.

96. Box 250, File 1-60, Completion Report for the Month of December, 1897.

97. Box 256, File 451-556, Report of Operations for the Month of March, 1900.

98. Box 283, Taylor to Wilson, November 2, 1900.

99. *ARCE,* 1900, Serial 4089, 1039; *ARCE,* 1901, Serial 4279, 902.

100. *ARCE,* 1902, Serial 4444, 784.

101. Box 276, File 3282, H. W. Jack to Millis, October 2, 1902.

102. Box 267, File 2545, F. V. Abbot to Millis, April 23, 1901.

103. Box 267, File 2545, Millis to Gillespie, August 10, 1901.

104. Ibid; Box 267, File 2545, Millis to Gillespie, September 10 and October 7, 1902.

105. Box 253, File 576-680, Completion Report for the Week Ending May 14, 1899.

106. "Accident at the Gravel Quarry," *Leader,* June 1, 1899.

107. Box 254, Pile 831-910, Preston to Walker, June 20, 1899.

108. Box 255, File 1-110, Preston to Walker, January 17, 1900.

109. Box 254, File 911-1000, Swigart to Walker, August 12, 1899.

110. Box 254, File 911-1000, Preston to Walker, June 20, 1899.

111. Box 254, File 911-1000, Preston to Walker, August 10, 1899.

112. Box 250, File 61-126, Joseph E. Kuhn to Taylor, January 25, 1898.

113. Box 250, File 61-126, Wilson to Taylor, April 3, 1898. It wasn't necessary to give Taylor that instruction since the platforms were normally finished in advance of other work.

114. Box 250, File 127-180, Wilson to Taylor, April 22, 1898. The contractors did not favor extending the work to a double shift of eight hours each instead of the single shift of 10 hours. Both firms thought that it would cost more: 15 percent additional at Admiralty Head, while the Pacific Bridge Company claimed that the extra labor would cost them twice as much as provided for in its exist-ing contract. Philip Eastwick did not support the expansion of the project under Maney, Goerig and Rydstrom since there was no indication that they could manage a larger operation with any more competency than they could a single shift. Taylor seems to have

intervened directly with each firm to establish a double shift over the objections of the contractors. Box 293, File 526-560, Pacific Bridge Company to Taylor, May 3, 1898 and Maney to Taylor, May 4, 1898.

115. Box 250, File 61-126, Kuhn to Taylor, January 25, 1898.

116. Box 250, File 61-126, unsigned letter, April 3, 1898.

117. Box 250, File 347-400, Ricksecker to Taylor, August 23, 1898.

118. Box 250, File 127-180, Wilson to Taylor, April 25, 1898.

119. Box 250, File 127-180, Wilson to Taylor, May 3, 1898.

120. Matthew L. Adams, *Designating U. S. Seacoast Fortifications* (privately printed, 2001), liv–lviii; Adams communication to the author, August 28, 2000.

121. The naming of the various batteries is detailed in the appendix.

122. Box 281, File 3205, Annual Inspection of Fort Flagler, June 3-10, 1904.

123. *Island County Times*, July 17 and 31, 1903.

Chapter 3

1. "Ryan-Hitchcock Marine Fortification," 41st Congress, 2nd Session, House Executive Document 17.

2. Ibid., 47.

3. Admiralty Head: Box 291, File 53-85, Joseph E. Kuhn to Taylor, August 14, 1897; Fort Worden: Box 251, File 471-550, C. H. McKinstry to Taylor, September 8, 1898. As constructed, the battery at Key West showed no apparent influence of the Fort Worden design.

4. San Francisco: Box 297, File 980-1649, R. H. Ober to Millis, August 19, 1903; Columbia River; Box 304, File 370.1, Chittenden to Lt. Col. S.W. Roessler, June 30, 1908.

5. Despite such encouragement at the district level, designs still had to be reviewed in the Office of the Chief of Engineers. After several reviews there, the Chief of Engineers, with the assistance of the Chief of Artillery when that position was created, examined them again. If they were found satisfactory, the plans went to the War Department for final approval and were only then returned to the originator for execution.

6. Winslow, *Notes on Seacoast*, 110.

7. Ibid., 111.

8. Box 283, File 2513, Taylor to Col. Charles R. Suter, June 4, 1898.

9. Ibid.

10. Ibid., Suter to Taylor, June 14, 1898.

11. Ibid., Taylor to Suter, July 2, 1898.

12. Ibid., Millis to Gillespie, November 10, 1902.

13. Ibid., Taylor to Suter, July 2, 1898.

14. Ibid., Millis to Gillespie, November 10, 1902.

15. Box 288, File 3203, Completion Report for the Fiscal Year Ending June 30, 1904.

16. Box 283, File 2553, Commanding Officer Fort Worden to the Adjutant General, June 30, 1902.

17. Box 269, File 3205, Preston to Chittenden, November 5, 1906. Because the alteration began at different times on different emplacements, the work was completed in a staggered sequence. All the work was completed by the end of 1906, but armament was not reinstalled in Batteries Ash and Quarles until 1907.

18. Winslow, *Notes on Seacoast*, 199.

19. Ibid., 201.

20. Ibid., 207.

21. The formal dialogue between the board and Honeycutt is contained in the article "Sea Coast Mortar Fire," cited above, and "A Reply to the Report of the Board on Sea Coast Mortar Fire," *Journal of the United States Artillery* [hereafter cited as *JUSA*], Volume 8, No. 2 (September-October 1897). A mathematical discussion of Honeycutt's theories appears in the same issue of the *Journal* under the title "The Probability of Hit When the Probable Error in Aim is Known."

22. *Handbook of American Coast Artillery Materiel*, Ordnance Department Document No. 2042 (Washington, D.C.: GPO, 1923), 350.

23. Box 257, File 827-952, Capt. J. D. C. Hoskins to the Adjutant General, Report of Inspection for the Quarter Ending June 30, 1900.

24. Box 272, File 2970, Mason to Millis, August 13, 1903.

25. Box 275, File 3205, Annual Inspection of Fort Casey, June 12-15, 1904, by Maj. John P. Wisser.

26. The next mortar battery to be constructed following that at Fort Casey was to have been at Fort Flagler. However, the construction plant at Flagler was not in order at the time Taylor was directed to begin plans for the design, and the Fort Worden plant was

functioning. As a result, construction of the battery was shifted to Worden. Box 253, File 437-500, Taylor to Wilson, April 6, 1899.

27. The concrete used in the batteries at Puget Sound was made from imported European cement. No cement was manufactured in the northwest, and transportation costs made the price of eastern cement prohibitively high. Imported cement was admitted duty-free since it was to be used by the government, but each shipment required a special dispensation from the Secretary of War. Congress passed the appropriation bills with the proviso that only products of American manufacture be used except where the quality of the defenses demanded otherwise. The continuing need for concrete caused a Board of Engineers in 1903 to examine the cements manufactured on the west coast. It found that even at that date, there was not sufficient experience to indicate the true quality of the product. The board endorsed foreign cements since their performance was well known and thus they were used in the defenses to the apparent exclusion of all others. See Box 254, File 1001-1093, Taylor to Wilson, August 19, 1899; Box 247, File 1601-1674, Gillespie to Millis, August 8, 1901; Box 242, File 2640 #2, Reports of a Board of Engineers, November 3, 1903.

28. Box 253, File 576-680, Completion Report for the Week Ending May 14, 1899.

29. Winslow, *Notes on Seacoast*, 238.

30. Box 273, File 3040; an extensive correspondence concerning the correction in August 1903.

31. "Battery Wilhelm, Fort Flagler, Washington," U.S. Engineer Office, Seattle, September 6, 1929, File No. F/13/7/18, Washington State Parks and Recreation Commission.

32. Box 285, File 2903, Ober to Millis, September 5, 1903.

33. Box 299a, File 345.1, Mason to J. B. Cavanaugh, January 26, 1916. The same treatment was applied to Battery Lee.

34. Box 242, File 2675, Mason to Millis, February 14, 1905.

35. Box 243, File 3040, Millis to A. McKenzie, February 10, 1905.

36. Winslow, *Notes on Seacoast*, 242.

37. Ibid., 252.

38. Doubtless this room within a room was the inspiration behind the lining of the Fort Casey batteries mentioned earlier.

39. Winslow, *Notes on Seacoast*, 169.

40. Box 283, File 2553, Alfred Mordecai to the Chief of Ordnance, August 23, 1902.

41. Ibid., 88.

42. RG77, ammunition hoist correspondence, Greble: Lloyd England to Adjutant, Artillery District of Narragansett, September 9, 1903; New York: J. M. K. Davis to Adjutant General, Department of the East, August 31, 1903; endorsement of G. L. Gillespie to the letter of Charles R. Suter, October 7, 1903.

43. England to Adjutant, September 9, 1903.

44. RG 77, ammunition hoist correspondence, D. E. Hughes to Edgar Jadwin, April 4, 1902.

45. Box 254, File 1151-1233, Taylor to Wilson, September 2, 1899. The use of a differential block was not new. When coupled with a crane it had served as the primary ammunition delivery system in two-story emplacements built before 1896. The crane hoist continued as an emergency delivery system in all later gun batteries even though the Taylor-Raymond hoist had proven itself entirely adequate and dependable. The emergency ammunition service was more necessary with the balanced platform hoist. Should a lift fail, Taylor foresaw difficulties in moving the heavy projectiles within the emplacements whose ceiling trolleys had been designed to move ammunition toward the hoist shafts and not in any other direction. To partially compensate for the deficiency, he installed light railroad tracks in the shot gallery floors of the Fort Casey and Fort Flagler main batteries. The tracks led to the exterior where they emerged adjacent to the crane. He envisioned that small cars, each capable of carrying a single projectile, would run on the rails and carry shot and shell to the outside where they would be picked up by the crane and hoisted to the loading platform. The cars were designed but not built and the system, used in the Columbia River forts as well, was never fully completed.

46. Winslow, *Notes on Seacoast*, 90.

47. Box 241, File 2509, F. V. Abbot to Millis, June 21, 1904.

48. Box 267, File 2512, H. J. M. Baker to Chittenden, December 3, 1906; Box 281, File 3201, Report of Operations at Fort Flagler, December 1906.

49. Box 269, File 3205, Inspection of Fort Ward, January 18, 1904, G. S. Grimes, Commanding Artillery District of Puget Sound.

50. Box 283, File 2513, Preston to Chittenden, September 27, 1906.

51. Box 271, File 2570, Mason to Millis, January 6, 1903; Box 277, File 2506, Millis to Gillespie, October 30, 1902.

52. Box 321, File 330.5, Mason to Kutz, July 6, 1910.

53. Box 281, File 3205; Annual Inspection of Fort Flagler by Maj. John P. Wisser, June 3-10, 1904.

Chapter 4

1. Endicott Board, 21-25.

2. *Annual Report of the Chief of Ordnance* [hereafter cited as *ARCO*], 1886, Serial 2465, 20.

3. *ARCO,* 1894, Serial 3302, 37.

4. *American Coast Artillery Materiel, 222.*

5. *ARCO,* 1890, Serial 2836, 28.

6. "Annual Report of the Principal Operations of the Watertown Arsenal," Appendix 14, *ARCO,* 1895, Serial 3378, 2.

7. "Report of Principal Operations at the Watertown Arsenal," Appendix 28, *ARCO,* 1892, Serial 3083, 419.

8. Ibid.

9. Maj. James N. Williams, "Advance in Coast Artillery Guns and Carriages from 1892 to 1912," *JUSA,* Volume 38, No. 3 (November-December 1912), 2807; Lt. C. C. Gallup, "Development and Construction of Modern Gun Carriages for Heavy Artillery, *JUSA,* Volume 4, No. 1 (January 1895), 34.

10. Lt. George D. Squier, "The International Electrical Congress of 1893, and its Artillery Lessons," *JUSA,* Volume 3, No. 1 (January 1894), 8.

11. Lt. M. F. Harmon, "The Buffington-Crozier Experimental Disappearing Carriage for 8-in. Breech Loading Steel Rifle," *JUSA,* Volume 4, No. 1 (January 1895), 46-54.

12. Ibid., 55. RG77, Defenses of the Columbia River, F. W. Phisterer to the Adjutant, Artillery District of the Columbia, June 7, 1906. After the tests, the experimental model was sent to Fort Columbia and emplaced in Battery Ord, and there its behavior changed considerably. By 1906, it could only be made to raise from the loading position by beating on it with a heavy sledge, and after firing, it stubbornly refused to "disappear." Officers decreed it to be "useless for any purpose whatever." The carriage was later replaced.

13. Box 249, File 23, Taylor to the Chief of Engineers, November 30, 1896.

14. Box 254, File 831-910, Joseph E. Kuhn to Taylor, July 21, 1899.

15. Box 271, File 2505, Chief of Engineers to Taylor, July 13, 1900.

16. Box 272, File 2970, James Benet to F. A. Pope, February 12, 1906.

17. File 272, File 2915, F. V. Abbot to Millis, April 25, 1901.

18. Box 265, File 725-870, Report of Operations at Fort Casey for the Fiscal Year Ending June 30, 1902.

19. Box 241, File 2509, Chittenden to McKenzie, May 7, 1906.

20. Ibid.

21. Box 273, File K, Report of Capt. J. D. C. Hoskins, July 1, 1901.

22. Box 281, File 3205, July 31, 1904, response of post ordnance officer to inspection of June 3-10, 1904.

23. Box 244, File 3205, Annual Inspection of the Artillery District of Puget Sound, October 10-25, 1905.

24. Box 279, File 2971, Mason to Pope, March 16, 1906. The cause of the accident was not certain. Since the shape of the emplacement restricted the movement of the gun, it may have been that the carriage was traversed too far, crushing the loading platform against the traverse wall. The likelihood of such an accident was recognized in 1902, when it was proposed that the rigid loading platform be replaced with a collapsible grid of hinged bars that would operate like a roll-top desk; the platform could be abbreviated when the gun was at extreme traverse. The alternative was to remove parts of the traverse wall. Predictably, District Engineer John Millis preferred altering the carriage over an action which in his opinion would reduce the effectiveness of the emplacement. The Ordnance Department was adamant about the alteration of any coast defense weapon, and it is not surprising that it was the emplacements that were modified to allow a somewhat greater traverse. Box 283, File 2553, Millis to Chief of Engineers, August 14, 1902; Box 277, File 2506, Millis to Gillespie, October 30, 1902; and Box 279, File 2971, Millis to Gillespie, November 25, 1902.

25. Box 302, File 358.1, Report of Inspection, June 2-5, 1907.

26. Box 323, File 355.6, Preston to Kutz, February 16, 1909.

27. Lt. William C. Harrison, "The Belcher Slow-Motion Elevating Device for 12-Inch Barbette-Guns," *JUSA,* Volume 43, No. 1 (January-February 1915), 106.

28. H. L. Hawthorne, "The Naval Attack on Sea-Coast Fortifications," *JUSA,* Volume 6, No. 1 (July-August 1896), 8.

29. Winslow, *Notes on Seacoast,* 407.

30. Ibid., 411.

31. Box 254, File 1094-1150, "Outline Project for the Mine Defense of Puget Sound," September 8, 1899. The mine case was changed by inserting a corrugated steel cylinder between the halves of the

standard spherical mine case. The effect was to increase the volume of the case and thus its buoyancy.

32. "Torpedo Defense of Rich's Passage Entrance to the Port Orchard Naval Station," September 8, 1899; drawing in author's collection.

33. Ibid.

34. Winslow, *Notes on Seacoast*, 412.

35. Results of an inspection made by Maj. James P. Wisser, October 24, 1905. Wisser's inventory reflects the same number of mines as was included in the 1899 program: 229.

36. Box 252, File 181-215, Taylor to Wilson, January 18, 1899.

37. Box 253, File 751-830, Eastwick to Taylor, July 5, 1899.

38. Box 252, File 181-215, Taylor to Wilson, January 18, 1899.

39. Box 257, File 689-770, Annual Report for the Year Ending June 30, 1900.

40. Box 242, File 2631, Millis to Gillespie, April 9, 1903. Agate Pass, west of Bainbridge Island, formed another approach to the Naval Station although it could not be negotiated by heavy warships. A casemate, thirty-six mines, and two batteries of rapid fire guns were to have formed the defenses. Land was acquired for the casemate but its construction was deferred until time of war. Proposals to build the batteries persisted until 1915 and then faded. The mines themselves, stored at Middle Point, were the only visible evidence of the intended barrier at Agate Pass. Box 298, File 322.1, endorsement of F. V. Abbot to letter of Chittenden to H. J. M. Baker, August 17, 1907.

41. Box 269, File 3205, Report of Inspection by Maj. John P. Wisser, October 24, 1905.

42. Box 244, File 3171, "Remarks of a Board Convened at the Artillery District of Puget Sound to Review Mine Defense at Rich's Passage," September 21, 1905.

43. Ibid., endorsement of November 21, 1905.

44. Box 320, File 32.4, Chief of Engineers to Chittenden, December 17, 1906.

45. Box 320, File 355.4, Chief of Coast Artillery to the Adjutant General, November 20, 1908. Battery Mitchell was briefly considered again ten years later when it was thought that it might be useful as a defense against submarines. RG 392, Entry 231, letter from the Coast Defense Ordnance Officer to the Commanding Officer, Coast Defenses of Puget Sound, July 5, 1918.

46. Site Plan of Fort Ward Military Reservation, May 1915, author's collection.

47. Box 244, File 3205, G. F. Humphrey to Military Secretary, October 18, 1905.

48. *Annual Report of the Chief of Coast Artillery* [hereafter cited as *ARCCA*] 1909, Serial 5717, 308.

49. Ibid., 305.

50. Capt. Ernest A. Greenough, "Planting and Raising Mines From Scow," *JUSA*, Volume 46, No. 1 (July-August 1916), 60 and 62.

51. Box 305, File 39.8.1, Report on Preparedness, Pacific Coast Artillery District, June 30, 1913.

52. NARA, RG 156, Box 1118, File 12727, Wilson to Maj. Gen. Nelson Miles, April 25, 1898.

53. Interview with Homer Reynolds, December 2, 1963. Reynolds served as the caretaker of Fort Whitman from 1925 until it was abandoned in 1944.

54. "Report of Major J. P. Story, 7th Artillery," *JUSA*, Volume 12, Nos. 2 and 3 (September-October, November-December, 1899), 183.

55. Box 250, File 181-260, Transfer receipt, May 14, 1898. The depression position finder had problems of its own. It was a touchy instrument and observers had to be highly skilled in placing the horizontal crosshair of the telescope precisely at the waterline of the vessel being tracked. It was also sensitive to refraction, the bending of light rays by differences in the density of the atmosphere. Some of the effects of refraction could be compensated for by adjusting the instrument to read the known ranges to markers called datums.

   Baselines, whether horizontal or vertical, were established only for batteries of heavy guns. Cannon below six inches in caliber were aimed with the carriage sight alone. Exceptions were rare. Battery Tolles had a horizontal baseline because of its critical position on the western approach to Admiralty Inlet. Here, accurate fire at the longest ranges possible was a necessity.

   Early target practice with three-inch and five-inch guns was very poor and, after World War I, the three-inch batteries in Puget Sound were supplied with coincidence range finders, which had been used experimentally in other defenses since 1909. The range finders were optical instruments consisting of a long horizontal tube, with lenses at opposite ends of the tube that projected separate images of the target onto a central eyepiece; the action of bringing the separate images into coincidence actuated a mechanical range indicator, and hence the device was called a coincidence range finder. It was the same principle as the horizontal base system, except that the baseline was restricted to the length of the single instrument. There were two types in Puget Sound, the nine-foot and the fifteen-foot instrument. A small concrete range-finder station was built near each three-inch battery in 1918 and 1919. The stations for Batteries Thornburgh and Vinton were turned over to the Artillery after the armament for those batteries had been removed. *ARCCA,* 1909, Serial 5717, 303; Box 301, File 348.1, Preston to the Chief of Engineers, September 4, 1918; Box 320, File 348.8.

56. Box 241, File 2554, Adjutant General's report of engagement, October 7, 1903; Box 244, File 3104, J. P. Story to Chief of Engineers, August 4, 1904.

57. Maj. Harold E. Cloke, "The Development of Coast Artillery Target Practice," *JUSA,* Vol. 39, No. 1 (January-February 1913), 43.

58. This was the Wadsworth system, so called because of its origins at that post in New York Harbor. It established the basic principles upon which the Barrancas system was founded. Garland Whistler figured as prominently in the creation of the Wadsworth system as he did in the subsequent Pensacola tests.

59. Ibid., 36-40.

60. Annual Report of the Chief of Artillery, 1903, Serial 4629, 368 [hereafter cited as *ARCA*].

61. *Annual Report of the Secretary of War*, 1907, Serial 5271, 29 [hereafter cited as *ARSW*].

62. *ARCA*, 1906, Serial 5106, 220.

63. Box 244, File 3104, Report of the Pratt Board, October 26, 1904.

64. Box 244, File 3199, Chief of Artillery to the Commanding Officer, Artillery District of Puget Sound, December 2, 1905.

65. Ibid.

66. Box 243, File 3104, James Allen to the Chief of Artillery, May 19, 1906.

67. Box 243, File 3104, F. V. Abbot to Chittenden, October 26, 1906.

68. Pensacola Report, 45.

69. Box 243, File 3104, Capt. John Stephen Sewell to A. MacKenzie, August 16, 1905.

70. Winslow, *Notes on Seacoast*, 345. For a detailed exposition of the origin of Sewell construction, see Nelson H. Lawry, "Fixed in Concrete: Sewell Construction in Tactical Buildings," *Coast Defense Study Group Journal*, Volume 10, Issue 4 (November, 1996), 30-37.

71. Box 301, File 348.1, "Position Finding Equipment, Artillery District of Puget Sound," December 24, 1910.

72. The cost of the installation is estimated at about $739,000. The figure is based on the estimate included in the Taft Board Report, less the amount for batteries not constructed. Even at this rate, the total sum of the installation was exceeded only by that of San Francisco.

73. "Proposed Range Towers, Location E," Drawer 103 Sheet 107-3, RG 77, NARA.

74. "Fire Control System, Fort Casey, Wash.," 1 March 1908, in the collection of the Washington State Parks and Recreation Commission.

75. "Fort Flagler, Washington," November, 1907, site plan in the author's collection.

76. Box 244, File 3104, Ober to Millis, April 10, 1905.

77. *ARCA*, 1906, 221.

78. Box 281, File 3205, Annual Inspection of Fort Flagler, June 3-10, 1904.

79. Pensacola Report, 13.

80. Winslow, *Notes on Seacoast*, 324.

81. Box 301, File 348.1, construction estimates.

82. "BC Benson-Ft. Worden, Wash.," Drawer 103, Sheet 72-61, RG 77, NARA. The station had been scheduled for replacement at least since 1922.

83. Box 313, File 660.2, Cavanaugh to Chief of Engineers, February 20, 1915.

84. Box 317, File 348.3, H.E. Cloke to Commanding Officer, Coast Defenses of Puget Sound, January 28, 1916.

85. *Installation and Maintenance of Fire Control Systems at Seacoast Fortifications* War Department Document 483 (Washington: GPO, 1915), 13-16 and 287-92.

86. Pensacola Report, 50.

87. Ibid.

88. Lt. William Lassiter, "Range and Position Finding," *JUSA,* Volume 4, No. 2 (April 1895), 242.

89. Box 276, File 3313, Mason to Chittenden, September 20, 1906.

90. Capt. William C. Davis, "A Proposed Automatic Sight for Disappearing Guns," *JUSA,* Volume 21, No. 2 (March-April 1904), 190.

91. Box 301, File 348.1, Circular letter of the Chief of Engineers, May 29, 1912.

92. Box 288, File 3203, Completion Report for the Fiscal Year Ending June 30, 1905.

93. Winslow, *Notes on Seacoast*, 358.

94. Maj. L. B. Bender, "Transmission of Data for Heavy Artillery," *JUSA*, Volume 55, No. 5 (November 1921), 419.

95. Box 323, File 348.6, modification correspondence, March 1923.

96. *ARCCA*, 1917, Serial 7341, 936-637.

97. Box 305, File 348.1, Completion Report, March 1918.

98. Box 300, File 345.1, C. L. Sturdevant to the Chief of Engineers, August 29, 1919.

99. The underground switchboards were dismal places. No shred of sunlight entered, nor was there adequate ventilation, and in 1920, the post surgeon at Fort Casey commented on the unhealthful conditions. He had treated one of the two operators for recurrent headaches caused by bad air and eyestrain. He recommended that the administrative lines (those not involved in fire control) be moved to a better room in the post headquarters; the casemated switchboard could be maintained with weekly use. The change was made and other posts followed suit. NARA, RG94, "Records of the Adjutant General's Office, Fort Casey, 1917-1939," letter of post surgeon to fort commander, February 10, 1920.

Chapter 5

1. *ARSW*, 1906, Serial 5105, 44. As of June 30, 1906, all but four of the 1,300 gun and mortar emplacements in the national defenses were complete. A few large caliber guns and a little less than half of the rapid-fire armament remained to be mounted.

2. Capt. James F. Howell, "Guns for the Defense of the Outer Harbor," *JUSA*, Volume 23, No. 2 (March-April, 1905), 117.

3. Box 283, File 250.8, Millis to Mackenzie, April 11, 1905. Further refinements came in 1909 at the behest of Garland Whistler, commander of the defense. His idea recognized some of the peculiarities of the main battery design. He proposed assigning two guns only to Battery Randol and making Battery Quarles a three-gun battery. Since the third emplacement of the newly organized Battery Quarles was connected to Battery Ash through interior passageways, Whistler advocated the removal of Quarles' third ten-inch gun and its replacement by a twelve-inch weapon. The emplacement would then be made part of Battery Ash. The third emplacement of Battery Quarles had no interior communication with the first and second emplacements, and if no twelve-inch gun could be found for the third emplacement, the existing ten-inch gun was to be placed out of commission, to be held only for war services. Such an arrangement, Whistler argued, would use the main battery to its best natural advantage, and allow for more efficient administration, service practice, and instruction. His proposal was accepted, and since there was no twelve-inch gun to replace the existing armament in Battery Quarles, the third ten-inch gun of that battery was placed out of commission. Box 313, File 660.2, Garland Whistler to the Adjutant General, January 29, 1909; and the sixth endorsement, March 9, 1909.

   The problem is illustrated in Plate 4-6. The gun is pointed northwest, away from the entrance to Admiralty Inlet and toward the Strait of Juan de Fuca. As one result, the crew is bunched awkwardly to one side. Another result is that the gun carriage now blocks access to the emergency ammunition service, essentially a grate-covered opening in the emplacement floor. That may explain the activity at the railing, where men appear to be lifting projectiles with a jerry-rigged tackle, demonstrating perhaps another means of service should the mechanical ammunition hoist cease to function during an engagement.

4. Box 300, File 345.1, Mason to C. L. Sturdevant, August 27, 1919.

5. Box 244, File 320.5, Annual Inspection of the Artillery District of Puget Sound, June 2-15, 1904.

6. Box 313, File 660.2, Chittenden to Mackenzie, November 23, 1907.

7. Box 244, File 3205, Annual Inspection of the Artillery District of Puget Sound, October 10-25, 1905.

8. Box 313, File 660.2, J. B. Cavanaugh to the Commander, Department of the Columbia, February 7, 1912.

9. apps.westpointaog.org/Memorials/Article/2879, accessed December 20, 2013.

10. U. S. Army Corps of Engineers, Seattle District, "Puget Sound Project for Defense," Millis to Gillespie, March 29, 1902.

11. Box 313, File 660.2, Chittenden to Mackenzie, November 23, 1907.

12. Box 272, File 2915, Millis to Gillespie, July 7, 1902; Box 244, File 2170, miscellaneous correspondence, Millis to Gillespie and Mackenzie, 1902-1904.

13. Editorial in the Boston *Globe* as quoted in "Our Coast Defense," *The National Magazine,* Volume IV, No. 2 (May, 1896), 115.

14. Lt. Samuel E. Allen, "Trained Artillery for the Defense of Sea-Coast Forts," *JUSA,* Volume 4, No. 2 (April, 1896), 225.

15. Taft Board Report, 23.

16. Box 244, File 2170, Abbot to Millis, February 24, 1905.

17. The Board deleted defenses for the Penobscot River and Port Royal. Taft Board Report, 12.

18. Ibid., 23.

19. Box 313, File 660.2, Chittenden to Mackenzie, November 23, 1907.

20. Maj. H. L. Hawthorne, "Increased Calibers with Low Velocities versus Smaller Calibers with High Velocities for Seacoast Guns," *JUSA,* Volume 28, No. 2 (September-October 1907), 111.

21. Capt. Herman W. Schull, "The 12-inch Gun versus the 14-inch Gun," *JUSA* Volume 29, No. 2, (March-April 1908), 116.

22. Ibid. Discussions of the various advantages claimed for the twelve-inch gun and the fourteen-inch gun were heated and the adoption of the fourteen-inch gun was by no means a quiet event. It was in large measure a contest between ordnance officers, who preferred guns capable of firing projectiles of great striking energy and large explosive charge, features embodied in the 14-inch gun; and artillery officers, who advocated weapons with the principal qualities of accuracy and rapidity of fire, attributes cited for the twelve-inch gun. Hawthorne, "The 12-inch Gun versus the 14-inch Gun," *JUSA,* Volume 30, No. 1 (July-August 1908), 5.

23. Maj. Edward P. O'Hern, "Guns, Ammunition, and Accessories," *JUSA,* Volume 40, No. 2 (September-October, 1913), 137.

24. Taft Board Report, 11.

25. *ARCCA,* 1916, Serial 7140, 795.

26. Box 283, File 2553, Grimes to Secretary of War, October 21, 1904.

27. Box 283, File 2553, Grimes to Military Secretary, May 3, 1905 and November 13, 1905.

28. Box 299, File 330.1, "Construction of Batteries, October 1908-January 1909."

29. Ibid.

30. Box 322, File 330.6, Kutz to the Chief of Engineers, September 18, 1908.

31. Box 244, File 3205, Annual Inspection of the Artillery District of Puget Sound, October 10-25, 1905.

32. Box 321, File 3205, Kutz to the Chief of Engineers, October 8 and November 7, 1908. A ten-inch gun battery also was to be part of the Deception Pass defenses, the guns to have come from the demolished Battery Benson. When the proposed demolition was dropped, the guns were no longer available, and the ten-inch battery was not a part of active discussions at the time when construction plans were being formulated. The proposed battery remained an element of the approved plan until 1911 when it was finally stricken because of the small size of Goat Island. Box 313, File 660.2, Chief of Engineers to the Adjutant General, October 11, 1911.

33. Winslow, *Notes on Seacoast,* 369-70.

34. Ibid., 370.

35. Box 299a, File 341.1, "Searchlight Project Report," October 11, 1917.

36. While a range equal to that of the guns was the goal, the searchlights installed subsequent to the Taft Board were disappointing. Under favorable conditions, a white ship could be illuminated by a sixty-inch projector sufficiently well for tracking at not more than 8,000 yards. If the same vessel were darkly colored, the range dropped to about 5,500 yards. Hines and Ward, *The Service of Coast Artillery,* 406.

37. Winslow, *Notes on Seacoast,* 373.

38. Ibid., 394.

39. Maj. W. C. Davis, "The Location and Tactical Employment of Searchlights in Coast Defense," *JUSA,* Volume 31, No. 3 (May-June 1909), 248.

40. Searchlight Project Report.

41. Ibid.

42. Ibid.

43. Chittenden to Mackenzie, November 23, 1907.

44. Ibid.

45. Box 299a, File 34.1, Chief of Engineers to C. W. Kutz, August 23, 1910.

46. *Annual Report of the Chief of Staff,* 1914, Serial 6798, 127.

47. *ARCCA,* 1916, Serial 7140, 1174.

48. Ibid.

49. Box 313, File 660.2, notations dated December 14, 1915, and prepared by Capt. Arthur R. Ehrnbeck.

50. Box 313, File 660.2, Cavanaugh to Chief of Engineers, April 29, 1916.

51. Box 313, File 660.2, A. H. Acher to Chief of Engineers, October 11, 1917.

52. "Status of Seacoast Batteries," 31 December 1943, tabulation in the files of the Office of the District Engineer, Seattle. Other six-inch and sixteen-inch gun batteries were planned, but those mentioned here were the only ones completed.

53. "Table of United States Army Cannon and Projectiles," Ordnance pamphlet 1676, Washington: GPO, 1915. The figure does not include sub-caliber cannon used in target practice, fourteen-inch cannon identified with the Taft Board, or the 1895 sixteen-inch type gun.

54. Capt. Beverly W. Dunn, "Projectiles, Fuzes and Primers," *JUSA,* Volume 20, No. 1 (July-August 1903), 26.

55. Box 313, File 660.2, 1917 memorandum of E. J. Dent, otherwise undated.

56. *ARCCA,* 1917, Serial 7341, 943.

57. "History of the Northwestern Sector," Enclosure No. 1, 10.

58. Memorandum for the Chief of Ordnance from the Chief of Artillery, July 19, 1907, included in Coast Artillery School materials, Washington State Parks and Recreation Commission.

59. Chittenden to Mackenzie, November 23, 1907.

60. *ARCCA,* 1915, Serial 6967, 788.

Chapter 6

1. Box 278, File 2695, Ober to Millis, December 19, 1903.

2. Box 322, File 337.6, Mason to Cavanaugh, May 24, 1913.

3. Box 316, File 386.2, Mason to Kutz, April 1, 1910.

4. "General Plan, Emplacements 1&2; 3&4; 5, 6&7," June 23, 1903, Drawing 103-22-12 through 14, RG77, NARA.

5. Box 280, File 3175, Taylor to Hoskins, November 8, 1900.

6. Box 323, File 342.6, Kutz to Preston, May 21, 1910. The incident appears to be associated with the closure of the main gate in 1910, ordered by Whistler in a running disagreement with the City of Port Townsend regarding the safety of the road adjacent to Fort Worden. For the same reason he also denied access to a boy delivering milk to the post.

Garland Nelson Whistler was irascible and colorful. He was a member of a distinguished military family and distantly related to the artist James McNeil Whistler. How he received his commission is obscure. At the request of Garland's father, President Lincoln in 1864 asked Secretary of War Edwin Stanton to admit the young Whistler to West Point, but for whatever reason he was not enrolled. President Andrew Johnson appointed him a second lieutenant in 1867, apparently without prior military training or experience.

His last duty assignment was as commander of the Puget Sound defenses, where he arrived in 1908, and by that time he had made significant contributions to the development of smokeless powder and the apparatus of coast artillery fire control. He was considered a "close student of history" and was a frequent lecturer on many topics—the military of course, but also Henry VIII and the Masonic Order, of which he was an enthusiastic member. He also gave the occasional piano concert for his friends, preferring to play in his stocking feet.

His reputation as a figure of note preceded him to the Northwest where, at the laying of the cornerstone for the armory in Bellingham, an observer described him as a "striking figure" and a military genius, adding that as "befits a genius, it pleases him to wear his hair long." Almost everyone commented on his hair, a shrubby mass that seemed to escape from under his service cap like errant steel wool. He reached mandatory retirement age in 1911 and died several years later. *Collected Works of Abraham Lincoln,* vol. 7, University of Michigan, letter to Edwin M. Stanton, August 20, 1864; *Papers of Ulysses Simpson Grant,* vol. 16, Southern Illinois University Press, letter to Andrew Johnson, October 9, 1866; Horace W. McCurdy interview and Louis H. Hansen interview, Jefferson County Historical Society Oral History Project, 1989; "Col. Whistler Will Lecture This Evening," *Leader,* March 3, 1910; "Colonel Whistler Explains Closing of Fort Gates," *Leader,* July 27, 1910; "Washington Considers Importance of Reviving the National Guard," *Oregonian* (Portland), November 20, 1910; "Whistler Will Go East," *Oregonian* (Portland) April 7, 1911.

7. Box 279, File 3025, Ober to Millis, July 29, 1903.

8. Box 246, File 3320, F. V. Abbot to Millis, November 15, 1902.

9. Box 246, File 7624, F. V. Abbot to Chittenden, August 30, 1906.

10. Box 307½, File 660, Chittenden to Mason, August 13, 1907.

11. Maj. W. G. Haan, "Coast Artillery Reserves," *JUSA,* Volume 33, No. 3 (May-June 1910), 303.

12. "History of the Northwest Sector," 184.

13. Diary of E. H. Sargent, Acting Assistant Surgeon, Fort Casey, entry for November 19, 1899; copy in the author's collection.

14. Ibid., November 14 and November 22, 1899.

15. Ibid., October 7, 1899.

16. *Leader,* February 26, 1904; February 1, 1910.

17. *Leader,* March 1 and 12, 1912.

18. *Leader,* January 23, 1903; December 10, 1903; March 18, 1906; January 11, 1912; February 21, 1912; March 2, 1917; and *Island County Times,* November 18, 1921; July 7, 1922.

19. Box 295, File 866-900, Preston to Walker, September 26, 1898; *ARCA,* 1906, Serial 5106, 204; *Leader,* June 16, 1915.

20. *Leader,* June 3, 1904; February 24, 1910; September 6, 1911.

21. Lt. Col. J. R. J. Jocelyn, R. A., in "Professional Notes," *JUSA,* Volume 3, No. 3 (July 1894), 494.

22. Maj. John Ruckman, "Seacoast Artillery Practice," *JUSA,* Volume 30, No. 2 (September-October 1908), 160.

23. Box 241, File 2554, Commanding Officer Artillery District of Puget Sound to the Adjutant, Department of the Columbia, October 7, 1903.

24. Box 313, File 660.2, Whistler to Adjutant General, September 10, 1910.

25. Glenn Rowley letter, December 9, 1917, Interpretive Services Collection, Washington State Parks and Recreation Commission. Experts in coast defense will question that part of the letter that has only two men ramming a heavy shell into the breech of the gun; the letter was meant for the writer's girlfriend, so perhaps some allowance should be made for a lapse in technical accuracy.

26. Maj. James M. Williams, "The Development of Coast Artillery Gunnery in the United States During the Last Twenty Years," *JUSA,* Volume 42, No. 3 (November-December 1914), 285-87.

27. Maj. W. B. Hardigg, "Shore Batteries vs Battleships," *JUSA,* Volume 55, No. 1 (July 1921), 45.

28. Capt. Edward Canfield Jr., in "Correspondence," *JUSA,* Volume 34, No. 3 (November-December 1910), 351-54.

29. *ARCA,* 1906, Serial 5106, 219.

30. *ARSW,* 1909, Serial 5716, 43.

31. "Standing of Companies of the Coast Artillery Corps, Based on Figures of Merit Attained in Service Practice, 1913," War Department, Bulletin No. 44, Sept. 28, 1914.

32. *ARSW,* 1908, Serial 5420, 28.

33. Lt. Samuel E. Allen, "Trained Artillery for the Defense of Sea-Coast Forts," *JUSA,* Volume 4, No. 2 (April 1895), 226.

34. Maj. Gen. J. P. Sanger, "First Steps Toward a System of Target Practice and Fire Control for Sea-Coast Artillery," *JUSA,* Volume 49, No. 3 (May-August 1918), 134.

35. Carey Sanger, "The Organization of Our Sea-Coast Artillery Forces," *JUSA,* Volume 5, No. 1 (January 1896), 17; Lt. John Ruckman, "Artillery Difficulties in the Next War," *JUSA,* Volume 2, No. 3 (July 1893), 430.

36. *Annual Report of the Lieutenant-General Commanding the Army,* 1901, Serial 4271, 50 and 89.

37. General Order 25, Headquarters of the Army, Adjutant General's Office, February 28, 1909.

38. Letter of H. C. Davis in "Correspondence," *JUSA,* Volume 31, No. 1 (January-February 1909), 107.

39. *ARSW,* 1905, Serial 4942, 25.

40. *ARCA,* 1906, Serial 5106, 210.

41. *Annual Report of the Chief of Staff,* 1907, Serial 5271, 179. The 150th Company was assigned to the Fort Ward mine defense. The 149th Company was assigned to Fort Casey, an action probably related to the nearness of the post to the minefield at Fort Whitman.

42. "History of the Northwest Sector," 178-85.

43. "Report of the Artillery Inspector, Department of the East," *JUSA,* Volume 16, No. 3 (November-December 1901), 324.

44. *Annual Report of the Adjutant General,* 1912, Serial 6378, 452. It was rumored that there was a colony of deserters living in British Columbia, although the existence of such a colony was never demonstrated.

45. *Annual Report of the Department of the Columbia,* 1905, Serial 4944, 223.

46. Box 241, File 2554, Commanding Officer, Artillery District of Puget Sound to the Adjutant, Department of the Columbia, October 7, 1903. See also *ARSW,* 1916, Serial 7140, 182, for an identification of some of the vocations within the coast artillery.

47. *ARCCA,* 1909, Serial 5717, 256.

48. *ARCA,* 1903, Serial 4629, 362.

49. *ARCA,* 1904, Serial 4782, 414.

50. *ARCA,* 1906, Serial 5106, Table A facing 230; *ARSW,* 1907, Serial 5271, 8 and 9.

51. *ARCA,* 1906, Serial 5106, Table A.

52. "Artillery Difficulties," 423; "Letters on Sea-Coast Artillery," *JUSA,* Volume 3, No. 3 (July 1894), 362.

53. E. M. Weaver, "An Experiment with Militia in Heavy Artillery Work," *JUSA,* Volume 7, No. 1 (January-February 1897), 1-15.

54. *ARSW,* 1902, Serial 4443, 39.

55. Weaver, "An Experiment with Militia," 11.

56. Box 248, File 1-1685, James A. Drain to Millis, July 13, 1903.

57. David A. Maurier, July 24, 1961, personal communication to author.

58. *ARCA,* 1907, Serial 5272, 208.

59. *ARCA,* 1908, Serial 5421, 224.

60. *ARCA,* 1907, 193.

61. *ARSW,* 1909, Serial 5716, 65.

62. Washington National Guard Pamphlet 870-1-5, "The Official History of the Washington National Guard," in six volumes, Headquarters, Military Department, State of Washington, Office of the Adjutant General, Camp Murray, Wash., Volume 5, 458.

63. *ARCCA,* 1909, 264-265.

64. *ARCA,* 1908, 224. This number rose and fell depending upon the proposals to add to the defenses; in 1909 it was estimated that the Washington militia would have to provide 96 officers and 2,270 enlisted men. By 1915, a landmark year, the number had been revised to 67 officers and 1,894 enlisted men. Lt. Col. W. Irving Taylor, CAC, NGNY, "National Guard Coast Artillery," *JUSA,* Volume 48, Nos. 2 and 3 (September-December 1917), 191.

65. "History of the Washington National Guard," Volume 5, 463.

66. The 1st and 3rd Companies, Coast Artillery Reserve Corps, were mustered out in 1914 because of their minimal strength. They were later reestablished. Ibid., 464.

67. *ARCCA,* 1912, Serial 6378, 975.

68. The figure is an average based on the reports of the Judge-Advocate General, 1912-1916.

69. *ARCCA,* 1915, Serial 6967, 786.

70. *Annual Report of the Chief of Staff,* 1916, Serial 7140, 164 and 181.

71. "History of the Northwest Sector," 149-50.

72. Ibid., 189 and 194. The other regiments were the 69th, 65th, 49th, 48th, 39th, and 24th Artillery, CAC.

73. "History of the Washington National Guard," Volume 6, 604.

74. "History of the Northwest Sector," 178-82.

75. Box 286, File 3078, Millis to Mackenzie, February 27, 1904.

76. Lt. Frank S. Clark, "A War Condition Period and How It Can Be Best Utilized," *JUSA,* Volume 4, No. 3 (May-June 1914), 259.

77. "Coast Artillery Fire Instruction," *JUSA,* Volume 3 No. 2 (April 1894), Lt. Henry J. Reilly, 215 and Lt. John W. Ruckman, 276. The possibility of land attack was first treated in official form in the 1898 Coast Artillery Drill Regulations.

78. Box 313, File 660.2, Lt. C. R. Pettis to Chittenden, May 1, 1908.

79. Box 244, File 320.5, Annual Inspection of the Artillery District of Puget Sound, June 2-15, 1904, Maj. J. P. Wisser.

80. Box 27a, File 3015, Ober to Millis, October 30, 1903.

81. *ARCA,* 1907, Serial 5272, 208.

82. "Department of Artillery and Land Defense, Coast Artillery School: Advanced Course, 1910 to 1911," a collection of several loose-leaf binders containing notes and materials from several of the instruction sequences from the Coast Artillery School at Fort Monroe, Virginia; Interpretive Services Collection, Washington State Parks and Recreation Commission.

83. Box 313, File 660.2, Chittenden to Pettis, November 18, 1907.

84. Box 244, File 3104, Millis to Mackenzie, January 5, 1905.

85. Col. William C. Davis, "Additional Practices in the Coast Defenses of Pensacola," *JUSA,* Volume 46, No. 1 (July-August 1916), 52-59.

86. *ARCCA,* 1916, Serial 7140, 1174. There was a single piece of ordnance that was designed for early land defense within the fortifications. The small six-pounder rifle was mounted on a wheeled carriage and could be maneuvered into a number of prepared positions, where it was chained in place. It was cumbersome and unpopular, and its role was filled later by more conventional light field guns. None of the six-pounders were sent to Puget Sound.

87. "Editorials," *JUSA,* Volume 53, No. 1 (July 1920), 71.

Chapter 7

1. *Leader,* November 19 and 20, 1918.

2. The barbette carriage for ten-inch and twelve-inch guns could elevate to fifteen degrees. The disappearing carriage for ten-inch guns could elevate to twelve degrees and the same carriage for twelve-inch guns, the most powerful weapon in many harbors, could manage but ten degrees. *Table of United States Army Cannon and Projectiles,* Ordnance Pamphlet 1676, Washington: GPO 1915, 30-31.

3. Ordnance Department Document No. 2042, 110.

4. "Report on the Navy Test Firings at the Battleship San Marcos (Texas) as a Target, March 20-21, 1911," copy in the author's collection. [Hereafter cited as "San Marcos Report."]

5. *ARCCA,* 1912, Serial 6378, 979.

6. A single class of depression position finder could read ranges up to twenty thousand yards, but it could be used only at sites with an elevation greater than 280 feet and was hence not a real alternative. *Description of Lewis Depression Position Finders,* Ordnance Pamphlet 1876, Washington: GPO 1916, 7.

7. "San Marcos Report."

8. Capt. John W. Gulick, "Armor and Ships," *JUSA,* Volume 38, No. 3, (November-December, 1912), 276.

9. Allan R. Millett, Peter Maslowski, and William B. Feis, *For the Common Defense: A Military History of the United States from 1607 to 2012* (New York: Free Press, 2012), 283.

10. *ARCCA,* 1911, Serial 6197, 747.

11. *ARCCA,* 1919, Serial 7685, 4049.

12. Capt. Edwin Landon, "The Needs of the Coast Artillery," *JUSA,* Volume 25, No. 2 (March-April 1906), 147.

13. Box 320, File 355.4, unsigned bill of lading dated October 20, 1917.

14. Box 313, File 660.2, E. E. Winslow to District Engineer, March 18, 1919. The removal of some armament that was considered to be of "insufficient military value" had been under consideration at least since 1915. *ARSW,* 1916, Serial 7140, 182.

15. *ARCCA,* 1917, Serial 7341, 939.

16. Box 318, File 355.3, Preston to the Commanding Officer, North Pacific Coast Artillery District, February 20, 1918. The balanced pillar mechanism was removed from the carriage and the guns were emplaced as barbettes.

17. "General History of the Harbor Defenses of the Columbia, 1864-1945," Appendix C (Fortifications), July 28, 1945. Copy of typescript in the author's collection. The six-inch guns were taken from Grays Harbor in 1932 and apparently stored at Fort Worden until 1937 when they were emplaced as Tolles 'B.'

18. Box 320, File 355.4, Charles L. Phillips to the Commanding Officer, North Pacific Coast Artillery District, November 15, 1918, and endorsements.

19. *Annual Report of the Chief of Staff,* 1911, Serial 6197, 146.

20. *ARCCA,* 1911, Serial 6197, 747.

21. *ARCCA,* 1914, Serial 6798, 563-65.

22. "History of the Northwest Sector," 151.

23. *ARCCA,* 1917, Serial 7341, 942.

24. Major F. E. McCammon, "Future Seacoast Artillery," *JUSA,* Volume 54, No. 2, (February, 1921), 141.

25. Maj. Gen. Johnson Hagood, *We Can Defend America* (New York: Doubleday, Doran and Co., Inc., 1937), 47.

26. W. R. W. James, R. A., reprint from the *Journal of the Royal Artillery,* May 1910, in "Professional Notes," *JUSA,* Volume 34, No. 1 (July-August 1910), 66.

27. *ARCCA,* 1919, Serial 7685, 4043-48.

28. *ARCCA,* 1920, Serial 7806, 1906.

29. Maj. Paul D. Bunker, "Many Things," *JUSA,* Volume 54, No. 6 (June 1921), 579.

30. Lt. Meade Wildrick, "The Possibilities of Railroad Coast Artillery," *JUSA,* Volume 47, No. 2 (March-April 1917), 144.

31. *ARCCA,* 1920, Serial 7806, 1899.

32. Capt. F. E. McCammon, "Another Come-Back for Captain Christmas," *JUSA,* Volume 53, No. 3 (September 1920), 267.

33. "History of the Northwest Sector," Addenda, History of the 14th Coast Artillery, 1.

34. General Order 13, War Department, June 9, 1925.

35. "The Field Officers' Course, Coast Artillery School," *JUSA,* Volume 54, No. 3 (March 1921), 258-65.

36. Box 320, File 355.4, Ordnance Officer, Ninth Corps, to Ordnance Officer, Harbor Defenses of Puget Sound, August 5, 1925, and fifth endorsement, June 3, 1926.

37. David P. Kirchner and E. R. Lewis, "The Defenses of Puget Sound," unpublished ms in author's collection, 24.

38. "History of the Northwest Sector," Enclosure No. 1, 9. The anti-aircraft role seemed a natural for the Coast Artillery since both ships and airplanes were moving targets. An early indicator of the new duty was the discussion of the 1915 Board of Review to add anti-aircraft guns to existing fortifications. *ARSW*, 1916, Serial 7140, 182.

39. Lewis, *Seacoast Fortifications*, 102–103 and 110.

40. The airplane as a means of destruction had only limited bearing on the demise of pre–World War I fortifications protecting the harbors of the United States. Aircraft were a threat as observation posts for long range naval artillery, which was more to be feared than the frail and imperfect arsenal that might be fitted to the airplane of the 1920s and early 1930s. The warship had bested the defenses long before the advent of military air power.

41. Since 1898 it had been the practice to wash or paint the upper surfaces of batteries in a dull tone to harmonize with the landscape. Loading platforms and emplacement walls were usually colored to prevent glare. In Puget Sound, most of the batteries were high enough so that no specific camouflage effort was necessary, yet all the batteries were painted as a matter of course. In 1902 R. H. Ober compared the vegetation at Fort Flagler with colored papers and came up with a scheme, which was never implemented, for painting the batteries in varying shades of gray and green to reflect changes in the season. Most of the attempts at concealment were directed toward fire control stations. The observation towers were painted gray to fade into the background, and at Fort Casey, tree trunks were patterned across the first fire control buildings. Natural vegetation was encouraged, and some searchlight shelters were draped in chicken wire to provide a lattice for vines. Following World War I, the Chief of Engineers circulated recommendations for camouflaging seacoast batteries, based upon experiences in France. Under his proposal, emplacements were to have been masked with two shades of green and black. Nothing much seems to have been done until 1940, when an extensive program treated almost all the older Puget Sound batteries, covering them in nets woven with burlap, oznaburg, or chicken feathers. During this same period, many emplacements were painted in swirling patterns of green, rust, and yellow. Box 250, File 61-126, Joseph Kuhn to Taylor, May 2, 1898; Box 247, File 1676-2317, Ober to Millis, October 16, 1901; Box 279, File 3115, Ober to Millis, November 22, 1902; Box 299, File 327.1, mimeograph dated July 2, 1919; "History of the Northwest Sector," Enclosure No. 1, 9.

42. Box 313, File 660.2, Folder #3.

43. Lewis, *Seacoast Fortifications*, 124.

Epilogue

1. "Col. Kessler Dead at Fort Hancock," *New York Times*, September 16, 1935; obituary, *Sixty-seventh Annual Report of the Association of Graduates of the United States Military Academy at West Point, New York*, June 11, 1936, 196.

2. Eve Dravecky, email to the author, December 27, 2012.

3. Box 304, file 378.1, Winslow to district engineers, January 2, 1917.

4. Winslow, *Notes on Seacoast*, 6, 431, 438.

5. The content and quotations in this paragraph appeared in a slightly different form in the author's portion of the *Historic Fortification Preservation Handbook* (Olympia, WA: Washington State Parks and Recreation Commission, 2003), 2.6.

6. Browning, "Shielding the Republic," 266.

7. Samuel C. Florman, *The Civilized Engineer* (New York: St. Martin's Griffin, 1987), 42–48.

8. Florman, *The Existential Pleasures of Engineering* (New York: St. Martin's Griffin, 1996), 126.

9. Millet, et al., *For the Common Defense*, 282.

Index